Energy and Protein Requirements of Ruminants

Energy and Protein Requirements of Ruminants

An advisory manual prepared by the AFRC Technical Committee on Responses to Nutrients. Compiled by G. Alderman, in collaboration with B.R. Cottrill.

CAB INTERNATIONAL

CAB INTERNATIONAL
Wallingford
Oxon OX10 8DE
UK

Tel: +44 (0)1491 832111
Fax: +44 (0)1491 833508
E-mail: cabi@cabi.org
Telex: 847964 (COMAGG G)

A catalogue record for this book is available from the British Library.

ISBN 0 85198 851 2

First printed 1993
Reprinted with corrections 1995

Harvard-style references to this publication should be made as AFRC (1993)
with the full reference as:

AFRC (1993) *Energy and Protein Requirements of Ruminants*. An advisory manual
prepared by the AFRC Technical Committee on Responses to Nutrients.
CAB INTERNATIONAL, Wallingford, UK.

Printed and bound in the UK at the University Press, Cambridge

Contents

Chapter Two
Requirements for Metabolisable Energy   21

Chapter Three
Requirements for Metabolisable Protein   33

Chapter Four
Feed Evaluation and Diet Formulation   41

Chapter Eight
Goats

References

Appendix I
Tables of Feed Composition

Appendix II
Sequential List of Equations

Subject Index

List of Figures

List of Tables

Preface

The Agricultural Research Council's Technical Review "The Nutrient Requirements of Ruminant Livestock" published in 1980 proposed minor revisions to the ME system of calculating requirements, as detailed in their 1965 Technical Review, and proposed a new model for the calculation of the protein requirements of ruminants. Whilst the principles of the latter system received wide acceptance, use of the system in practice did not give satisfactory results. An advisory services' working party was set up in 1982 to review the system proposed, and to organise experiments to test any revised proposals. In 1983, the AFRC set up the Technical Committee on Responses to Nutrients (TCORN), which included representatives of the agricultural supply industry, Ministries and college advisers, as well as research scientists. This Committee produces technical reviews as seems appropriate, together with separate advisory publications.

The advisory services' Working Party on Energy Requirements, whose report was published in 1990 as AFRC TCORN Report No.5, "Nutritive Requirements of Ruminant Animals: Energy", recommended, with only minor alterations, the adoption of the AFRC 1980 Review's full recommendations on energy requirements of ruminant animals. No accompanying advisory manual was produced then, so this publication covers this topic as well.

The report of the Working Party on Protein Requirements has now been published as TCORN Report No.9, "Nutritive Requirements of Ruminant Animals: Protein". The Report recommends the adoption of a system based on Metabolisable Protein (MP) as the unit. This advisory manual is intended to assist in the adoption of the recommended MP system of calculating protein allowances for ruminants in the UK, replacing the Digestible Crude Protein system.

None of the ARC publications (1965;1980;1984) dealt with goats, but a TCORN Committee has been set up to review the "Nutrition of Goats". Material from the Committee's unpublished report has been included in this Manual at the express wish of the sponsors and with the agreement of the main TCORN Committee. Material from another AFRC TCORN Report No.8 on the "Voluntary Intake of Silage by Cattle" has also been included in the relevant sections of the text.

Preface to the Reprinted Edition

A number of errors in the figures and equations in the first printing of this Manual have been communicated to the compilers during 1994. The efforts of the correspondents concerned are gratefully acknowledged. The need for a reprint of the Manual has therefore created an opportunity to correct these errors, which did not affect the majority of the tables of requirements or diet examples.

The TCORN Working Party on Goat Nutrition revised its final manuscript in respect of the energy and protein contributions that can be expected from body reserves in lactating goats, thus affecting Tables 8.2 and 8.3 for dairy goats based on an earlier draft. Also, the proportion of true protein in the crude protein of goat milk used in calculating Table 8.1 was erroneously taken as 0.95 (as for cows) when the correct value for goat milk is 0.90 according to the TCORN Goat Nutrition Report. Tables 8.1, 8.2 and 8.3 have therefore been recalculated. Minor adjustments to the dairy goat diet examples were also needed.

Equations (76) and (78) concerning the ME equivalence of liveweight loss for lactating cattle and sheep were incorrectly stated in the text of the Manual on pages 32, 144 & 145, since division by k_l was omitted, but the ME requirement figures in Tables 5.1 and 7.1 have been confirmed as correct.

In the Tables of Feed Composition, the ME and FME values for maize gluten feed and rapeseed meal were not in accordance with ADAS (1994b) and the values have been updated. Minor corrections to the ME prediction equations for grass hays and dried lucerne have been made in accordance with ADAS (1994a).

November, 1994

Foreword

Since 1959 there has been coordinated action by public organisations to review periodically the literature on the nutrient requirements of farm livestock and update advice on feeding standards. Working parties of specialists drawn from research and advisory services and from the animal feed industry have collated, evaluated and interpreted the scientific information that is available worldwide.

This approach has a proven record of usefulness in producing over the years a succession of "state of the art" publications on "The Nutrient Requirements of Farm Livestock". These have attracted international recognition and provided an invaluable source of reference for those concerned with diet formulation, animal nutrition and advice to farmers on the feeding of livestock.

The present Technical Committee was set up in 1983 by the late Professor John Rook, to continue the role of the previous Agricultural Research Council Committee. Working Parties were established that included representatives of the research services, the universities, the advisory services and commerce. They were required to revise earlier publications and chart new opportunities for the development of feeding systems. A policy decision was taken to give more consideration to the role of nutrients as regulators of metabolism and to the dynamic relationship between nutrient intake and the responses of livestock. These Working Parties have also taken on responsibility both for the Technical Reports, explaining current understanding, highlighting gaps in knowledge and focusing research needs, and also for the preparation of practical recommendations on the feeding of livestock.

The form of publication that has been chosen for both Technical Reports and Advisory Booklets facilitates revision of particular subjects as and when sufficient new information becomes available. The Technical Reports are being published by CAB International in *Nutrition Abstracts and Reviews - Series B: Livestock Feeds and Feeding*, and the Advisory Booklets are published commercially. This Advisory Manual is the second to be published, the first being the *Nutrient Requirement of Sows and Boars*, which followed the publication of TCORN Report No.4 on the same subject in 1990.

This present Advisory Manual brings together the energy and protein requirements of ruminants, whose interdependence was highlighted by TCORN Report No.9 on the protein requirements of ruminants. It also implements in full in its tables of requirements (with the inclusion of agreed 5% safety margins), the recommendations of TCORN Report No.5 on the energy requirements of ruminants. The inclusion of numerous diet formulation examples is intended to assist the teaching and implementation of these integrated nutritional requirements for all classes of ruminants. The Acknowledgements that follow demonstrate the extensive collaboration of researchers and advisers, both from public and commercial institutions, who were involved in the preparation of this Manual.

Further reports are at an advanced stage of preparation with the following titles:

> Responses in the Yield of Milk Constituents to the Intake of Nutrients by Dairy Cows
>
> Nutrition of Goats

The Technical Committee and its Working Parties form a crucial link between basic research and the application of new knowledge in practice. This allows the public resources available for research to be utilised more effectively and enables the livestock industry to take information from research and incorporate it into feeding systems that make the best use of available resources.

Professor J.H.D. Prescott
Chairman of the Agricultural and Food Research Council's Technical Committee on Responses to Nutrients
Wye College, University of London

Acknowledgements

The drafting and checking of an Advisory Manual such as this, designed to give comprehensive, practical and up to date coverage of the AFRC reviews of the energy and protein requirements of ruminants, has involved numerous colleagues, whose contributions are hereby gratefully acknowledged. My chief collaborator was Dr B.R. Cottrill, ADAS, Starcross, who checked the derivation of the numerous equations required to generate the tables of requirements and advised on the layout of the Manual. Mr C.R. Savery, ADAS, Reading, assisted with the checking of the specially written software used to generate the numerous tables required. Detailed checking of all tables as set in the manuscript was undertaken by Dr J. Waters, (then with ADAS, Starcross, now with Trident Feeds) funded jointly (through UKASTA's Scientific Committee) by J. Bibby Agriculture Ltd, BOCM Pauls Ltd, British Sugar and Dalgety Agriculture Ltd.

Dr J.D. Sutton made available (with the permission of the Chairman of TCORN, Professor J.H.D. Prescott), the draft Report of the AFRC TCORN Working Party on "Nutrition of Goats". Dr D.I. Givens, ADAS, Stratford, advised on the Chapter on Feed Evaluation and the Tables of Feed Composition, whilst Dr C. Thomas, SAC, Ayr, supplied details of the prediction equations for forages and compound feeds now in use by ADAS, SAC, DANI and UKASTA.

Additional advice and comments on the manuscript were received from Dr D.M. Allen, Professor G.D. Barber, Dr A.T. Chamberlain, Dr M. Lewis, Dr N. Offer, Dr J.D. Oldham, Dr B.A. Stark and Professor A.J.F. Webster. Dr B.C. Cooke commented on behalf of the UKASTA Scientific Committee, in addition to releasing additional feed composition data on ruminant feeds.

The problems of copyright with such a publication as this were greatly eased by the offer of CAB International to publish the Advisory Manual, since they hold the copyright of most of the publications so extensively quoted herein. Their book publisher, Mr T. Hardwick, gave valuable advice on layout and typefaces.

G. Alderman,
Honorary Research Fellow
Department of Agriculture, University of Reading   December, 1993

Terminology and Symbols Used

This glossary of terms, symbols and units is comprehensive, in order that there is a logical and systematic use of abbreviations in the text of this Manual. The principles used in the construction of the glossary were:

UPPERCASE LETTERS are used for **energy and nutrient supply** (per animal) per day, either **g/d or MJ/d** as appropriate.

The same symbols, enclosed in square brackets, eg [DUP], are used for **concentrations, g/kg or MJ/kg,** except where existing usage dictates otherwise.

lower case letters are used for rates, efficiencies and proportions, which are either expressed as decimals (not percentages), or with relevant units, eg g/MJ.

The units and abbreviations used for weight, time etc are as in the SI system. Crude protein (CP) or (P) is taken as 6.25 x Nitrogen (N).

Subscripts are used to differentiate between metabolic functions as follows:

b	Basal metabolism
c	Concepta/gravid foetus/pregnancy
d	Dermal losses, scurf and hair
f	Liveweight gain
g	Gain/loss in liveweight in lactating animals
l	Lactation
m	Maintenance
n	Nitrogen utilisation, combined with the above set
o	Ovine
p	Production
t	Time in days
w	Wool/fibre growth

A   Activity allowance, J/kg/m or kJ/kg/d

a   Proportion of water soluble N in the total N of a feed

[ADIN]   Acid Detergent Insoluble Nitrogen in a feed, g/kgDM

B   Derived parameter in equation (14) to predict energy retention

b   Proportion of potentially degradable N other than water soluble N in the total N of a feed

BEN   Basal Endogenous Nitrogen, g/kg $W^{0.75}$/d

[BF], B%   Butterfat content of milk, g/kg or % per litre

C   Concentrate DM fed, kg/d

c   Fractional rumen degradation rate per hour of the *b* fraction of feed N with time, t

C1 - C6   Correction factors used in the calculation of the ME and MP requirements of ruminants

C_L Plane of nutrition correction factor in calculating ME requirements of lactating ruminants

[CDM]   Corrected Dry Matter, g/kg in grass silage only

CP, [CP]   Crude Protein, g/d in a diet or g/kgDM in a feed

DE, [DE]   Digestible Energy, MJ/d of a diet or MJ/kgDM in a feed

dg   Extent of degradation of feed nitrogen (or CP) at time, t

DMI, [DM]   Dry Matter, intake, kg/d, or content, g/kg in a feed

DMTP   Digestible Microbial True Protein, g/d, ie truly absorbed in the intestines (= Metabolisable Protein from microbes)

[DMTP]/[MTP] True absorbability of amino acids from Microbial True Protein

DOM, [DOMD] Digestible Organic Matter, kg/d in a diet or g/kgDM in a feed

dsi   True absorbability of amino acids derived from Undegradable Dietary Protein (UDP) as used in INRA (1988), see *dup*

DUP, [DUP]	Digestible Undegraded Protein (N x 6.25), the amount or proportion of undegraded feed protein that is truly absorbed, g/d in a diet or g/kgDM in a feed
dup	True absorbability of amino acids derived from Undegradable Dietary Protein, ie [DUP]/[UDP] (= *dsi* of INRA 1988)
E, [E]	Net Energy, MJ/d or g/kg
E_c	Net Energy retained in concepta, MJ/d
E_f	Net Energy retained in growing animal, MJ/d
E_g	Net Energy retained or lost in daily weight change in lactating ruminants, MJ/d
E_l	Net Energy secreted as milk, MJ/d
E_m	Net Energy for maintenance, MJ/d
E_{mp}	Net Energy for maintenance and production, MJ/d
E_t	Net Energy content of concepta at time t, MJ
E_w	Net Energy retained as wool or goat fibre, MJ/d
EBW	Empty-body weight, kg
[EE]	Ether Extract (oil), in feed, g/kgDM
ERDP, [ERDP]	Effective Rumen Degradable dietary Protein (N x 6.25), which has the potential to be captured by rumen microbes at a rumen digesta outflow rate of r/hour
$[EV_g]$	Energy Value of tissue lost or gained, MJ/kg
$[EV_l]$	Energy Value of milk, MJ/kg
exp	Exponential function using base *e*
F	Fasting metabolism, MJ/(kg fasted weight)$^{0.67}$
FE	Faeces energy, MJ/d
F_p	Proportion of forage in the diet Dry Matter

FME, [FME]   Fermentable ME of a diet, MJ/d or MJ/kgDM in a feed

GE, [GE]   Gross Energy of a diet, MJ/d or MJ/kgDM in a feed

HDMI   Hay Dry Matter Intake, kg/head/d

I   Intake of dietary ME, MJ/d scaled by fasting metabolism, F

[IVD]   *in vitro* digestibility [DOMD], g/kgDM of a feed

k   Derived parameter in equation (15) to predict energy retention

k_{aai} Efficiency with which a mixture of absorbed amino acids in ideal proportions is used for the net synthesis of protein as tissue, fibre or milk

k_c Efficiency of utilisation of ME for growth of the concepta

k_f Efficiency of utilisation of ME for weight gain

k_g Efficiency of utilisation of ME for weight change when lactating

k_l Efficiency of utilisation of ME for milk production

k_m Efficiency of utilisation of ME for maintenance

k_t Efficiency of utilisation of mobilised tissue for lactation

k_n Net efficiency of utilisation of absorbed amino acids, $= 1$ for maintenance and $= k_{aai}$ x RV for other purposes

k_{nb} Efficiency for basal metabolism (BEN)

k_{nc} Efficiency for growth of concepta (pregnancy)

k_{nd} Efficiency for synthesis of scurf and hair

k_{nf} Efficiency for growth

k_{ng} Efficiency for gain when lactating

k_{nl} Efficiency for lactation

k_{nm} Efficiency for maintenance

k_{nw}	Efficiency for wool and fibre growth
ln	Natural logarithm to base *e*
L	Level of feeding as a multiple of MJ of ME for maintenance
[La], La%	Lactose content of milk, g/kg or % per litre
LWG or ΔW	Liveweight gain or change, $\pm$ g or kg/d
M_c	ME requirement for growth of concepta, MJ/d
M_f	ME requirement for liveweight gain, MJ/d
M_g	ME requirement for liveweight change when lactating, MJ/d
M_l	ME requirement for milk production, MJ/d
M_m	ME requirement for maintenance, MJ/d
M_{mp}	ME requirement for maintenance and production, MJ/d
M_w	ME requirement for wool or fibre growth, MJ/d
M_E	Methane energy, MJ/d
[MADF]	Modified Acid Detergent Fibre in feed, g/kgDM
MCP, [MCP]	Microbial Crude Protein supply, g/d or g/kg
M/D	Metabolisable Energy, MJ/kgDM of a diet, see also [ME] for a feed or diet
ME, [ME]	Metabolisable Energy, MJ/d or MJ/kgDM of a feed or diet, see also M/D for a diet
[ME_{fat}]	Metabolisable Energy from fat (oil), MJ/kgDM in a feed
[ME_{ferm}]	Metabolisable Energy from fermentation acids, MJ/kgDM in a fermented or ensiled feed
MER	Metabolisable Energy requirement, MJ/d
MP, [MP]	Metabolisable Protein, g/d from a diet or g/kgDM of a feed
MP_c	MP requirement for growth of concepta, g/d

MP_f	MP requirement for liveweight gain, g/d
MP_g	MP requirement for liveweight gain when lactating, g/d
MP_l	MP requirement for milk production, g/d
MP_m	MP requirement for maintenance, g/d
MP_w	MP requirement for wool or fibre growth, g/d
MPR	Metabolisable Protein Requirement, g/d
MPS	Metabolisable Protein Supply, g/d
MTP, [MTP]	Microbial True Protein, g/d or g/kg
[MTP]/[MCP]	Proportion of Microbial Crude Protein present as True Protein
n	Lactation week number
$[N_a]$	Ammonia content of silage, g/kg total N
[NCD]	Neutral detergent cellulase [DOMD] in a feed, g/kgDM
[NCGD]	Neutral detergent cellulase + Gammanase [DOMD] in a feed, g/kgDM
NP_b	Net Protein equivalent of Basal Endogenous N, g/d
NP_c	Net Protein for growth of concepta (pregnancy), g/d
NP_d	Net Protein for scurf and hair growth, g/d
NP_f	Net Protein accreted in gain, g/d
NP_g	Net Protein accreted or mobilised when lactating, g/d
NP_l	Net Protein secreted in milk, g/d
NP_m	Net Protein for maintenance, g/d
NP_w	Net Protein for wool or fibre growth, g/d
NPN, [NPN]	Non-Protein Nitrogen, g/d of a diet or g/kg in a feed
[ODM]	Oven Dry Matter content of the fresh diet or diet, g/kg

[OMD]	Organic Matter Digestibility, g/kg of a diet or feed
[P], P%	Crude Protein content of milk, g/kg or % per litre
p	Effective degradability of feed N at outflow rate, r/h
q_m	Metabolisability of [GE] at maintenance, [ME]/[GE]
QDP, [QDP]	Quickly Degradable Protein (Nx6.25), g/d of a diet or g/kgDM in a feed
QDP/MCP	Limiting efficiency of conversion of QDP to MCP
R	Energy retention (E_r), MJ/d, scaled by fasting metabolism (F)
r	Rumen digesta fractional outflow rate per hour
RDP, [RDP]	Rumen Degradable Protein (Nx6.25), g/d in a diet or g/kgDM in a feed, for a given rumen outflow rate, r/h
RV	Relative Value of the amino acid mixture supplied, compared with the ideal amino acid mixture
SDMI	Silage Dry Matter Intake, kg/head/d
SDP, [SDP]	Slowly Degradable Protein (N x 6.25), g/d of a diet or g/kgDM in a feed, for a given outflow rate, r/h
SDP/MCP	Limiting efficiency of conversion of SDP to MCP
[TDM]	Toluene Dry Matter content of silage, g/kg
TDMI	Total Dry Matter Intake of a diet, kg/head/d
TP, [TP]	Tissue Protein (ARC 1980) = Net Protein, g/d or g/kg
u	Fraction of total feed N which is completely undegradable
UDP, [UDP]	Undegradable Dietary Protein (N x 6.25), g/d of a diet or g/kgDM in a feed, for a given outflow rate, r/h
UE	Urine Energy, MJ/d
w_n	Weight of feed (n) included in a diet, kgDM/d
W	Liveweight of the animal, kg

ΔW or LWG   Liveweight gain or change, $\pm$ g or kg/d

W_c Calf birthweight, kg

W_m Mature bodyweight of the dam, kg

W_n Liveweight of the animal in week n of lactation, kg

W_o Total weight of lambs at birth, kg

Y   Yield of milk, kg/d

y   Microbial protein yield in the rumen, gMCP/MJ of FME

Chapter One

Principles and Concepts

Metabolisable Energy

The ARC (1980) ME system is based on a basic relationship between the Metabolisable Energy (ME), intake from a feed or diet and the Net Energy (E) utilised or retained in the animal product, both expressed as MJ per day:

$$E = ME \times k \tag{1}$$

where k is the efficiency of utilisation of ME for the relevant metabolic process.

Definition of Metabolisable Energy

Metabolisable Energy (ME) intake is defined by ARC (1980) as the Gross Energy (GE), of the feed less that of the faeces (FE), urine (UE), and combustible gases (mostly methane, M_E), expressed as the SI unit Megajoules of ME per day (MJ/d) for a diet, or Megajoules per kilogram of feed or diet dry matter (MJ/kgDM). It represents that portion of the feed energy that can be utilised by the animal. ME is defined as:

$$ME = GE - FE - UE - M_E \tag{2}$$

The SI unit of energy used, the Megajoule, equal to 1,000 kilojoules (kJ), or 1,000,000 joules (J), can be converted to calories, the unit for heat used in France, The Netherlands and the USA, by using the exact conversion:

$$4.184 \text{ joules} = 1 \text{ calorie} \tag{3}$$

As the proportion of ME in the GE supplied by a diet declines as the level of feeding (L), increases, due to variation in the amounts of energy lost in faeces, urine and methane, ME measurements are defined as being measured at the maintenance level of feeding (L = 1).

Definition of Net Energy

The Net Energy of a feed or diet is that part of the digested feed energy that is utilised by the animal for bodily maintenance and production, after allowing for losses as faeces (FE), urine (UE), methane (M_E) and heat. The Net Energy content of an animal product is numerically the same as its content of Gross Energy, [GE], also known as Energy Value, [EV], expressed as MJ/kg. The [GE] of a feed or animal product is obtained by measuring the total heat evolved when the feed is combusted in oxygen in an adiabatic calorimeter. The Net Energy used by an animal for maintenance is equal to the heat production of the animal maintained at maintenance in a thermo-neutral environment.

Definition of Digestible Energy

Digestible Energy (DE) is defined as GE - FE, and as FE is the largest of the three terms to be subtracted, there is normally a good correlation between [DE] and [ME] values of feeds or diets, with [ME]/[DE] ranging from 0.81 to 0.86.

Metabolisability of feeds

The metabolisability of the [GE] of a feed at maintenance (q_m), is defined as the proportion of [ME] in the [GE] of the feed:

$$q_m = [ME]/[GE] \qquad (4)$$

An alternative term, the [ME] concentration in the feed or diet dry matter (M/D), was proposed by ARC (1965), because of the paucity of [GE] data on ruminant feeds. M/D is now well established as a method of calculating relevant values for k_m, k_f and k_l, when using the ME system in practice. However, AFRC (1990) recommended that greater precision would be achieved in diet formulation and performance prediction if q_m was more widely used as the basis for calculating the efficiencies of ME utilisation.

Conversion of q_m to dietary M/D values

Data on the [GE] content of ruminant feeds is now more widely available than previously, as MAFF (1990; 1992) quote [GE] values for all the feeds listed in their Tables. ARC (1965) and MAFF (1976) both assumed a mean [GE] value of 18.4MJ/kgDM (4.4Mcal/kgDM), in converting q_m values to M/D values. Since grass silage is predominant in most ruminant diets, it seems more appropriate to use a higher value for the average [GE] of ruminant diets, as grass silage has a mean [GE] of 19.0MJ/kg of Toluene Dry Matter (TDM) (MAFF 1992), whilst compound feeds have values varying from 18.4 to 18.8MJ/kgDM.

A mean value for the [GE] of ruminant diets of 18.8MJ/kgDM has therefore been used throughout this Manual when tabulating ME requirements. The appropriate q_m value used in the calculations is quoted in all cases.

Fermentable Metabolisable Energy

The Fermentable Metabolisable Energy, [FME], content of a feed or diet is required in order to use the Metabolisable Protein (MP) system of AFRC (1992). The unit [FME] was defined by AFRC (1992) as the ME content of the feed or diet, [ME], as MJ/kgDM, less the [ME] present as total oils and fats, $[ME_{fat}]$, and the ME contribution of fermentation acids, $[ME_{ferm}]$, present in partially fermented forages such as silage, or pre-fermented feeds such as brewery and distillery byproducts:

$$[FME] \ (MJ/kgDM) = [ME] - [ME_{fat}] - [ME_{ferm}] \qquad (5)$$

Dietary fats and oils, whilst highly digestible in the ruminant digestive tract, cannot supply molecules of energy yielding ATP to the rumen microbes. The average [ME] of fats and oils is of the order of 35MJ/kgDM, so that correction for the oil and fat content reduces the proportion of [FME] in the feed [ME] quite significantly, eg as little as 4% of oil reduces [FME] by 1.4MJ/kgDM.

The fermentation acids, primarily lactic, acetic, propionic and butyric acids, being themselves the end products of microbial fermentation, cannot yield any more energy (ATP) under the anaerobic fermentation conditions obtaining in the rumen, although they are efficiently absorbed and metabolised in the ruminant body. Their summated [ME] values (actually [GE] values in this case) are therefore deducted from the feed ME value. For example, 100g lactic acid/kg silage dry matter means a deduction of 1.51MJ from the estimated [ME] of the silage. Further details on the estimation of [FME] values are in Chapter Four.

Efficiencies of utilisation of ME

The efficiencies of utilisation of ME, k_m, k_f, and k_l as defined in ARC (1980), Appendix 3.II, p118, are defined by preferred linear equations involving q_m, the ratio of [ME] to the [GE] of a feed or diet:

Efficiency for maintenance $\qquad k_m = 0.35q_m + 0.503 \qquad (6)$

Efficiency for lactation $\qquad k_l = 0.35q_m + 0.420 \qquad (7)$

Efficiencies for growth,

$\qquad$ *growing ruminants* $\qquad k_f = 0.78q_m + 0.006 \qquad (8)$

$\qquad$ *lactating ruminants* $\qquad k_g = 0.95k_l \qquad (9)$

Additional equations for both k_m and k_f were proposed by ARC (1980) in Table 3.2, p80, for diets that were finely ground and pelleted, were all forage, or primarily first cut forages. AFRC (1990) tested these additional equations, but did not recommend their use for growing cattle, and they are not used here.

Values for other efficiency coefficients are assigned constant values, with no influence of q_m implied, namely:

Efficiency for growth of the concepta $k_c = 0.133$ (10)

Efficiency for utilisation of mobilised body tissue for lactation

$$k_t = 0.84 \qquad (11)$$

No efficiency values were stated for wool or fibre growth, so the assumption is that the efficiency for growth (k_f) should be used if the net energy retention in wool or fibre is a significant amount.

Correction for feeding level

The ME actually available to the animal, calculated as the sum of the ME (measured at maintenance) of the diet's component feeds, is reduced significantly at high levels of feeding due to the increased outflow rate from the rumen, and reduced retention time in the rumen. Two different approaches have been used by ARC (1980) to handle this problem.

Lactating ruminants

For dairy cattle, the decline is estimated by ARC (1980), p103, to be 1.8% per unit increase in feeding level (L) above maintenance ME requirement, excluding any safety margin. The correction factor C_L is calculated as follows:

$$C_L = 1 + 0.018(L - 1) \qquad (12)$$

where L is multiples of maintenance ME requirement.

AFRC (1990) extended the use of this correction factor to lactating ewes, and although AFRC (1993) did not do so with lactating goats, for the sake of consistency, the energy requirements of lactating goats have been calculated using this correction factor also.

Growing and fattening animals

Blaxter & Boyne (1970) derived an exponential function to describe the relationship between the energy retention (R), and the ME intake of growing cattle, in order to reduce the overestimation of energy retention implied by the linear model. The equation has the form:

$$R = B(1 - e^{-kI}) - 1 \qquad (13)$$

where   R is the retention of Net Energy,
and   I is the intake of dietary ME,
both scaled by the fasting metabolism of the animal (F).

The factors B and k are calculated from the efficiencies of utilisation of ME in equations (6) and (8) above, as follows:

$$B = k_m/(k_m - k_f) \qquad (14)$$

$$k = k_m \times \ln(k_m/k_f) \qquad (15)$$

It follows that both B and k are directly related to q_m, and can be tabulated, as in Table 1.1 overleaf. The curvilinear effects produced by equation (13) for different values of q_m are illustrated in Fig. 1.1, where the scaled ME intake, I (= ME/F), is plotted against the scaled energy retention, R (= E_f/F).

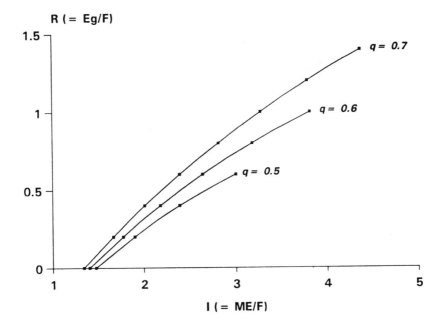

Fig. 1.1: Relationship between scaled energy retention (R) and ME intake (I).

Table 1.1: Relationship between q_m and the parameters B and k.

q_m	B	k
0.4	1.98	0.453
0.5	2.40	0.365
0.6	2.98	0.291
0.7	3.82	0.227
0.8	5.12	0.170

Calculation of ME requirements

The general equation for the calculation of ME requirements is given by rearranging equation (1) above:

$$ME \ (MJ/d) = E/k \qquad (16)$$

Lactating ruminants

For dairy cattle, lactating sheep and goats, this can be extended as follows:

$$M_{mp} \ (MJ/d) = C_L\{E_m/k_m + E_l/k_l + E_g/k_g + E_c/k_c\} \qquad (17)$$

Growing cattle and sheep

Equation (13) above is intended for performance prediction from known intakes of ME at a given value for q_m or M/D. However, the function can be rearranged (ARC 1980, p104), so as to calculate ME requirements for a given value of q_m:

$$M_{mp} \ (MJ/d) = (F/k) \ x \ ln\{B/(B - R - 1)\} \qquad (18)$$

B and k are calculated using equations (6), (8), (14) and (15) above, whilst scaled energy retention, R, for beef cattle is calculated using equations (40) for fasting metabolism (F), and equations (61) and (62) for [EV$_g$] (Chapter Two). For growing lambs, equation (41) for F, and (63), (64) or (65) are required for [EV$_g$], whilst equations (43) for F and (67) for [EV$_g$] are used for growing goats.

Principles of diet formulation

The ME system of ARC (1980) can be readily used to calculate ME requirements of a specified ruminant animal for a given value of q_m, or to predict performance from a given diet, since the total DM and ME intakes will be known, and the appropriate values for M/D or q_m can be calculated. With diet formulation, the ME required is influenced by the calculated M/D (q_m) of the diet being formulated. If the diet specification gives the M/D required, (or both DM and ME intakes), then two component diets of forage and compound can be formulated using:

$$M/D = x([ME] \text{ compound}) + (1 - x)([ME] \text{ forage})$$

which can be rearranged:

$$x = \frac{(M/D - [ME] \text{ forage})}{([ME] \text{ compound} - [ME] \text{ forage})} \tag{19}$$

where x is the decimal proportion of the compound in the diet.

The Variable Net Energy system for diet formulation

With more complex diets, achieving an exact fit to both DM and ME intake targets is more difficult. It also omits to consider diets that may be more economical at some other M/D value. If linear programming techniques are to be used in solving the problem, then diet formulation using appropriate feed Net Energy values, using the Edinburgh Variable Net Energy System (Harkins *et al.* 1974), was recommended by AFRC (1990), since Net Energy requirements (see Chapter Two) are independent of variations in M/D (q_m).

The procedure is simple in concept, if rather complicated to execute. The ME required for the specified level of animal production (M_{mp}), is calculated as specified earlier in this Chapter, using the metabolisability of the individual feed to do so, and using the appropriate level of feeding correction (L), or energy retention formula. Thus the [ME] of the single feed becomes M/D or defines the q_m of the diet, *ignoring DM intake constraints*. The E_{mp} requirement is also calculated in accordance with the equations in Chapter Two. *Both M_{mp} and E_{mp} are calculated without the addition of any safety margin.* The value for the joint efficiency of utilisation of ME for maintenance and production (k_{mp}) is then given by:

$$k_{mp} = E_{mp}/M_{mp} \tag{20}$$

The value for k_{mp} obtained is then used to calculate the Net Energy (E_{mp}) value for the feed when incorporated in the diet for the given level of animal production:

$$E_{mp} \text{ (MJ/d)} = M_{mp} \times k_{mp} \tag{21}$$

The calculated feed E_{mp} values can then be used in the conventional additive fashion to formulate a diet supplying the required amount of E_{mp}, *which should include the recommended 5% safety margin,* and taking cognisance of DM intake limitations. The example quoted by AFRC (1990) in their Appendix 1c (reproduced below) may clarify the steps in the calculations:

Growing beef cattle

Example   Calculation of Net Energies for maintenance and production

Feeds:	Hay	DM 860g/kg, 8.5MJ/kgDM ($q_m = 0.462$)
	Compound	DM 860g/kg, 13.0MJ/kgDM ($q_m = 0.700$)
Animal:	Liveweight	400kg
	Liveweight gain	0.65kg/d
	Male castrate of medium maturity	

E_m = 30.7MJ/d using equations (40) and (45)
E_f = 11.5MJ/d using equations (61) and (62) and Table 2.1
E_{mp} = 30.7 + 11.5 = **42.2MJ/d**

Scaled energy retention (R) for the hay with a q_m of 0.462 is:

k_m = 0.665 using equation   (6)
k_f = 0.366   "   "   (8)
k = 0.3971 "   "   (15)
B = 2.2241 "   "   (14)
E_g = 11.5   "   "   (61) and (62)
E_m = 30.7   "   "   (40)
R = 11.5/30.7 = **0.3746**

The M_{mp} requirement for hay alone is then calculated using equation (18):

$$M_{mp} \text{ (MJ/d)} = (E_m/k) \times \ln\{B/(B - R - 1)\} \qquad (18)$$

$$= (30.7/0.3971)\{2.2241/\ln(2.2241 - 0.3746 - 1\}$$

$$= \textbf{74.4}$$

Then k_{mp} = 42.2/74.4 = 0.567

and E_{mp} = 8.5 x 0.567 = **4.82MJ/kgDM**

Similar calculations for the compound give:

M_{mp} = 60.3MJ/d, k_{mp} = 0.700 and E_{mp} = **9.10MJ/kgDM**

These feed E_{mp} values can be used in any combination to achieve the required total requirement of E_{mp}, which with the addition of a 5% safety margin is 44.3MJ/d, so long as the total DMI is within appetite limits, as specified in Chapter Six.

Dairy cattle

Dairy cattle diets can be formulated using the same basic principles of calculating appropriate k_{mp} and E_{mp} values for the feeds selected, but using equation (17) to calculate M_{mp} of the feeds, and the appropriate Net Energy equations in Chapter Two to calculate the total E_{mp} required:

$$M_{mp} \text{ (MJ/d)} = C_L\{E_m/k_m + E_l/k_l + E_g/k_g + E_c/k_c\} \qquad (17)$$

Metabolisable Protein

Metabolisable Protein is defined as the total digestible true protein (amino acids) available to the animal for metabolism after digestion and absorption of the feed in the animal's digestive tract. Metabolisable Protein (MP), has two components:

1. *Digestible Microbial True Protein* (DMTP) produced by the activities of the rumen microbes, which synthesise protein from fermentable energy (FME) sources in the feed and amino acids or non-protein nitrogen from the breakdown of feed proteins in the rumen. About 0.25 of Microbial Crude Protein (MCP) is present as nucleic acids, which cannot be used by the ruminant for the synthesis of body tissue, milk etc. The Microbial True Protein (MTP) content is therefore 0.75 of the MCP. MTP is 0.85 digestible in the intestines, so that:

$$\text{DMTP (g/d)} = 0.75 \text{ x } 0.85 \text{ x MCP} = 0.6375\text{MCP (g/d)} \qquad (22)$$

2. *Digestible Undegraded feed Protein* (DUP), is that fraction of the feed which has not been degraded during its passage through the rumen (UDP), but which is sufficiently digestible to be absorbed in the lower intestines of the animal. The proportion of DUP in UDP varies from nil to 0.9, depending on the feed, its composition and pretreatment. The digestibility of UDP can be predicted from the Acid Detergent Insoluble Nitrogen [ADIN] content of the feed, or its Modified Acid Detergent Fibre [MADF] content (see p15).

Metabolisable Protein is therefore defined as:

$$\text{MP (g/d)} = 0.6375\text{MCP} + \text{DUP} \qquad (23)$$

The prediction of MCP supply from a given diet, and the estimation of DUP are dealt with later in this section.

Degradation of feed proteins in the rumen

The key protein parameters in the proposed MP system, QDP, SDP and DUP, are derived from measurements of the rates of degradation of feed proteins (dg)

suspended in a dacron bag in the rumen for various periods of time, normally up
to 48 hours for concentrates and 72 hours for forages (Ørskov & Mehrez 1977).
The proportion of the total nitrogen lost from the dacron bag with time is then
plotted against time (t), as shown in Fig. 1.2. The value at time zero is obtained
by washing the dacron and contents in a washing machine modified to give a
suitable cold rinse cycle. Similar data plots for typical concentrates are shown
in Fig. 1.3. An exponential function with three constants (a, b and c) is then
fitted to the data to give the curves shown in Figs 1.2 and 1.3. The equation has
the form:

$$dg = a + b\{1 - e^{(-ct)}\} \tag{24}$$

where a = *water soluble N extracted by cold water rinsing,*
       b = *potentially degradable N, other than water soluble N and*
       c = *fractional rate of degradation of feed N per hour.*

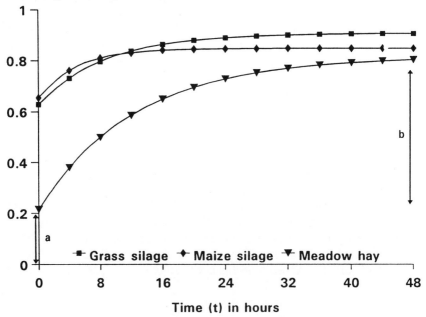

Degradability of total N

Fig. 1.2: N degradability of forages against time.

The high proportion of water soluble N in grass and maize silage should be
noted, and the slow rate of degradation of the remaining protein, compared to
air-dried hay. In Fig. 1.3 it can be seen that barley N degrades very rapidly in
the rumen, with little remaining after 12 hours, whilst at the other extreme, only
half of the protein N in fishmeal has been degraded after 24 hours, whilst
soyabean meal is intermediate. The time that feed stays in the rumen markedly
influences the proportion of the protein degraded by the microbes in the rumen.

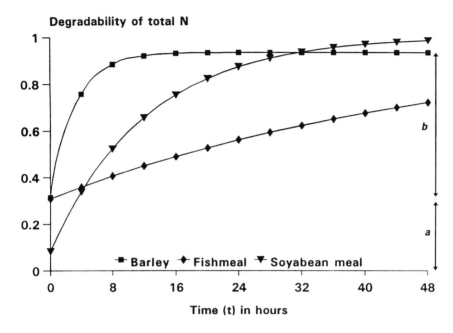

Fig. 1.3: N degradability of concentrates against time.

Rumen retention time and outflow rate

Retention time in the rumen is highly correlated with the level of feeding of the animal (L), since greater feed intakes result in faster outflow rates from the rumen. Measurements of fractional rumen outflow rates per hour (r), give values in the range 0.02 to 0.08/hour, implying that 0.02 to 0.08 of the total rumen contents would leave the rumen each hour. ARC (1984) gave estimates of suitable outflow rates in relation to level of feeding as follows:

1. Animals fed at a low level of feeding, about 1 x maintenance **0.02/h**
2. Calves, low yielding dairy cows (<15kg milk/d), beef cattle and sheep on higher levels of feeding, but less than 2 x maintenance **0.05/h**
3. High yielding dairy cows (>15kg milk/d), greater than 2 x maintenance **0.08/h**

A stepwise approach to this question inevitably results in problems as an animal's performance approaches one of the boundaries described above, eg a decline in milk yield from 16 to 14kg/d would imply that the outflow rate alters from 0.08 to 0.05, despite only a small change in feed intake. Although no relationship has been established experimentally, AFRC (1992) proposed the use of an empirical equation to smooth the operation of this factor, as follows:

$$r = -0.024 + 0.179\{1 - e^{(-0.278L)}\} \qquad (25)$$

This equation is not valid for feeding levels (L) less than 1. Outflow rates (r)
predicted from feeding level (L) using equation (25) are given in Table 1.2.

Table 1.2: Rumen outflow rate (r/h) as a function of level of feeding (L).

Level of feeding (L)	1.0	1.5	2.0	2.5	3.0	3.5	4.0	4.5
Outflow rate (r/h)	.019	.037	.052	.066	.077	.087	.096	.104

Influence of rumen outflow rate on degradability

Ørskov & McDonald (1979) showed that the effective degradability (p), for any
given rumen outflow rate (r), was given by:

$$p = a + (b \times c)/(c + r) \tag{26}$$

where a, b and c are the fitted constants from equation (24) above.

There is an associated mean rumen retention time for any given outflow rate,
shown in Fig. 1.4, together with the effects upon effective degradability.

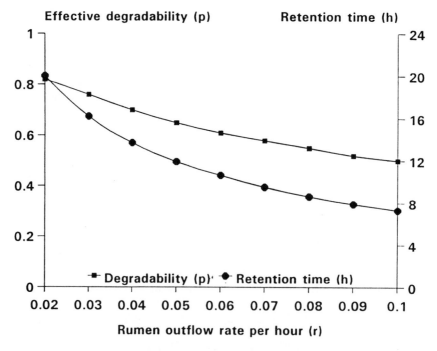

Fig. 1.4: Effect of rumen outflow rate on degradability and retention time.

The three fitted constants, (*a, b and c*) from equation (24) are combined with the outflow rate (r), appropriate for the level of feeding (L), of the animal to calculate the required protein fractions described by these functions.

Quickly Degradable Protein, [QDP]

The cold water extracted fraction of the feed total Crude Protein, [CP], defined by the constant, *a*, is called Quickly Degradable Protein, [QDP], and for any feed is calculated as:

$$[QDP] \ (g/kgDM) = a \times [CP] \ (g/kgDM) \tag{27}$$

This fraction of the total crude protein, comprising considerable amounts of non-protein N in the case of silages, but also water soluble small protein molecules, is released rapidly when the feed enters the rumen, resulting in an efficiency of capture by the rumen microbes of less than 1. Any urea added to the feed is to be included in the QDP fraction of the diet. Since ARC (1980) recommended that urea was only utilised in the rumen with an efficiency of 0.8, this efficiency has also been assigned to the [QDP] fraction of feeds.

The amount of QDP in a diet should be limited to less than 0.4 of the Effective Rumen Degradable Protein (ERDP), since ARC (1980), p169, suggest a limit on urea inclusion in diets to no more than 0.5g urea per kg liveweight (W), due to the risk of inducing ammonia poisoning. This is equivalent to 1.4g QDP per kg W, or 840g/d for a 600kg cow.

Slowly Degradable Protein, [SDP]

The amount of protein slowly degradable during the residence of the feed in the rumen is determined by the time spent in the rumen with the feed exposed to rumen bacterial digestion, which is a function of level of feeding (L) and outflow rate. The method of calculation is derived from equation (26) above:

$$[SDP] \ (g/kgDM) = \{(b \times c)/(c + r)\} \times [CP] \ (g/kgDM) \tag{28}$$

ARC (1980) assumed a *net* efficiency of capture for protein N of 1.0, which includes the effects of N recycling into the rumen from urea in blood and saliva.

Effective Rumen Degradable Protein, [ERDP] and ERDP

The term ERDP has been defined as a measure of the total N supply that is actually captured and utilised by the rumen microbes for growth and synthetic purposes. The [ERDP] content of a feed is defined as:

$$[ERDP] \ (g/kgDM) = 0.8[QDP] + [SDP] \tag{29}$$

The total ERDP supply to the rumen microbes (g/d), is then the sum of the weights of feed DM in kg/d (w_n) multiplied by the feed [$ERDP_n$] contents in g/kgDM, calculated as in equation (29):

$$ERDP \ (g/d) = w_1[ERDP_1] + w_2[ERDP_2] + w_3[ERDP_3] \ etc \qquad (30)$$

[ERDP] values of feeds bear no fixed relationship to ARC (1984) [RDP] values because of the variation in the [QDP] fraction of feeds, and they will be numerically smaller than [RDP] values at comparable outflow rates.

Digestible Undegradable Protein, [DUP] and DUP

ARC (1980) defined Undegradable Protein, [UDP], as Crude Protein, [CP], minus the Rumen Degradable Protein, [RDP]:

$$[UDP] \ (g/d) = [CP] - [RDP] \qquad (31)$$

With the partitioning of [RDP] into [QDP] and [SDP], [UDP] is now defined as:

$$[UDP] \ (g/d) = [CP] - \{[QDP] + [SDP]\} \qquad (32)$$

Instead of a constant digestibility of 0.85 for [UDP] in the lower intestines, the digestibility of [UDP] is now predicted from the Acid Detergent Insoluble Nitrogen, [ADIN], content of the feed (Goering & Van Soest 1970), as suggested by Webster *et al.* (1984), based on 37 raw materials used in compound feeds:

$$[DUP] \ (g/kgDM) = 0.9\{[UDP] - 6.25[ADIN]\} \qquad (33)$$

Table 1.3 gives values for a range of [UDP], [ADIN] and [MADF] values:

Table 1.3: [DUP] as a function of [UDP] and [ADIN] or [MADF] (g/kgDM).

		Undegradable Protein [UDP] (g/kgDM)											
ADIN	MADF	10	20	30	40	50	60	70	80	90	100	120	140
1.0	63	3	12	21	30	39	48	57	66	75	84	102	120
1.5	94	1	10	19	28	37	46	55	64	73	82	100	118
2.0	125	0	7	16	25	34	43	52	61	70	79	97	115
2.5	156	0	4	13	22	31	40	49	58	67	76	94	112
3.0	188	0	1	10	19	28	37	46	55	64	73	91	109
3.5	219	0	0	7	16	25	34	43	52	61	70	88	106
4.0	250	0	0	5	14	23	32	41	50	59	68	86	104

If [ADIN] data are not available, these authors suggested that [MADF]/62.5 could be substituted, *but this alternative should not be used for forages.* The equation gives a range of [UDP] digestibilities from 0.5 to 0.9, comparable to the INRA (1988) PDI system.

Effect of level of feeding (L) upon [ERDP] and [DUP] values of feeds

Since level of feeding (L) affects outflow rate (r), which in turn affects degradability as described by the Ørskov & McDonald (1979) equation (26) above and illustrated in Fig. 1.4, there are also effects of outflow rate upon [SDP], already defined in equation (28), and upon [DUP], via the effects of variation in [SDP] upon the calculation of [UDP] in equation (32). These significant effects are illustrated in Fig. 1.5 for soyabean meal, where a doubling of the [DUP] content occurs as L increases from 1 to 3 times maintenance, and a significant decrease in [ERDP] value occurs.

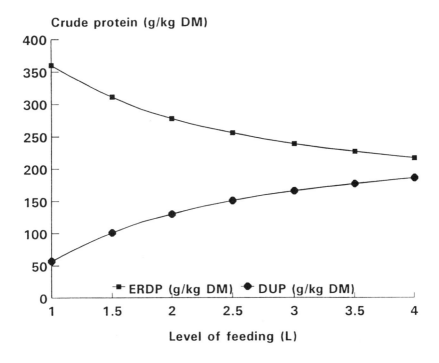

Fig. 1.5: Effect of level of feeding (L) on [ERDP] and [DUP] of soyabean meal.

Microbial protein synthesis in the rumen

Numerous feed and animal factors affect the level of microbial synthesis in the rumen, and they will not be comprehensively reviewed here. It is known that the type and amount of nutrients available from the diet and the synchronisation of nutrient release in the rumen affect the level of microbial synthesis, as also do growth promoters of both the antibiotic and ionophore type. It is not possible to quantify these effects on microbial protein synthesis adequately at present, but the MP system is sufficiently flexible for this information to be incorporated in due course, because of the factorial nature of the model proposed, as shown in Fig. 1.6. The factors taken into account in the MP system are:

1. Energy supply to the microbes.
2. Nitrogen supply to the microbes.
3. Level of feeding of the animal.
4. Outflow rate, principally determined by level of feeding.

1. *Energy supply to rumen microbes,* by which is meant those nutrients which, on microbial fermentation, yield the energy substrate adenosine triphosphate (ATP), necessary as a fuel to drive the rumen microbes' synthetic processes. *Energy supply is normally the first limiting factor on microbial protein synthesis.*

ARC (1980;1984) proposed the use of the ME supply from the diet as the determinant of microbial synthesis, based on a relationship with the amount of dietary organic matter digested in the rumen, about 0.65 of total DOM of the diet. AFRC (1992) concluded that moving towards rumen *fermentable* organic matter was desirable, rather than the term digestible organic matter, as was done by the French, Dutch and Nordic countries in their recently introduced systems. AFRC (1992) therefore proposed that a new unit, *Fermentable Metabolisable Energy,* FME, MJ/d or [FME], MJ/kgDM should be used for this purpose, as defined earlier in this Chapter, and estimated as shown in Chapter Four.

Accordingly, the rumen Microbial Crude Protein (MCP), yield (y), is expressed as gMCP per MJ of FME in the diet, and the level of feeding (L), determines the value of "y" for any particular situation. ARC (1984) proposed three levels of outflow rate as defined in the section on "*Rumen retention time and outflow rate*" above, which were essentially related to L as multiples of maintenance, so the latter basis was adopted by AFRC (1992) for the recommended levels of Microbial Crude Protein synthesis (y, gMCP/MJ of FME):

Values of Microbial Crude Protein synthesis (y)

All animals at maintenance level of feeding (L = 1)   9g MCP/MJ of FME
Growing sheep and cattle   (L = 2)   10g MCP/MJ of FME
Late pregnancy or lactating ewes and lactating dairy cows
                                             (L = 3)   11g MCP/MJ of FME

As discussed in the section on rumen outflow rates, such a stepwise approach results in problems in practice as an animal's performance moves from one defined level of feeding to another. AFRC (1992) proposed the use of another empirical equation to smooth the operation of this factor, but which adheres closely to their proposals for microbial yield, as follows:

$$y \text{ (g/MJ FME)} = 7.0 + 6.0\{1 - e^{(-0.35L)}\} \tag{34}$$

Values generated by equation (34) are given in Table 1.4.

Table 1.4: Microbial protein yield (g/MJ FME) as a function of level of feeding.

Level of feeding (L)	1.0	1.5	2.0	2.5	3.0	3.5	4.0	4.5
MCP yield (y)	8.8	9.5	10.0	10.5	10.9	11.2	11.5	11.8

As the level of feeding (L) plays such an important part in determining the probable level of microbial protein synthesis, the appropriate value for L is quoted in the tables of ME and MP requirements located in subsequent Chapters. Users of the MP system may choose to modify the value for "y" they use in the light of their knowledge of a particular diet or a feed additive.

2. *Nitrogen supply to microbes* may be limiting microbial protein synthesis more often than is realised, particularly if the ARC (1980;1984) proposals for RDP supply are relied upon. The AFRC (1992) MP system increases the amounts of ERDP required to match the amount of FME supplied by the diet and utilisable by the rumen microbes. By definition, ERDP (g/d) is a measure of the total N x 6.25 supply captured by the microbes, whether as non-protein N and/or intact soluble protein in the QDP, or degraded protein moieties, SDP. Also, by definition, ERDP is used with an efficiency of 1.0 for microbial protein synthesis MCP (g/d), *so we can state that in ERDP limiting situations:*

$$\text{MCP (g/d)} = \text{ERDP (g/d)} \tag{35}$$

Estimation of Microbial Crude Protein supply

MCP supply is calculated from the total FME in the diet, and the appropriate value for microbial protein yield (y), as defined by equation (34) and Table 1.4 above, provided that ERDP supply is greater than or equal to the MCP supply:

$$\text{MCP (g/d)} (\leq \text{ERDP}) = \text{FME (MJ/d)} \times y \text{ (gMCP/MJ FME)} \tag{36}$$

The ERDP supply required is also given by equation (36), since MCP cannot be greater than ERDP supply. The value for ERDP obtained from equation (36) must then be compared with the calculated total ERDP supplied by the diet. There are then three possibilities:

1. *If ERDP supply is less than ERDP requirement,* then the diet is ERDP limited and MCP supply is given by equation (35).

2. *If ERDP supply exceeds ERDP requirement,* then FME is first limiting MCP supply and equation (36) is valid. Excess ERDP will be wasted, and result in elevated levels of blood ammonia and urea.

3. *ERDP supply exactly matches the supply of FME.* This should be the objective of diet formulation using the MP system, because it avoids both unnecessary surplus N excretion, which has environmental importance on livestock farms, or else limitation of forage/diet intake caused by a shortage of ERDP. Advisors may choose to reduce ERDP supply, if a reduction in feed intake and milk yield is required, by reasons of milk quota limitation.

This test as to whether the estimated FME or ERDP supply from the diet is limiting MCP production is central to the calculation of MP supply.

Other protein systems deal with this problem differently. INRA (1988) defines two protein values for feeds. The first, [PDIE], is dependent upon the supply of energy as fermentable organic matter (FOM), whilst the other, [PDIN], is dependent on the supply of RDP. The Nordic system (Madsen 1985) also uses a unit, [AAT], based on the [FOM] of the feed and one based on the surplus/deficit of [RDP] required to match the [FOM], known as "Protein Balance in the Rumen", [PBV]. The AFRC (1992) proposals are based on the belief that it is easier to select feeds for balancing a diet if their [FME], [ERDP] and [DUP] contents can be seen in the feed tables, and that the calculation of MCP supply is only valid for a diet, rather than a single feed.

Estimation of Digestible True Protein supply

Having estimated Microbial *Crude* Protein supply, corrections are now required to estimate the amounts of *true* protein (amino acids) that will be absorbed in the lower digestive tract of the animal. This requires estimates of two fractions:

1. *True protein content of MCP (MTP), is estimated to be 0.75 of MCP* by AFRC (1992), compared to 0.80 suggested by ARC (1980;1984). Other authors have suggested values of 0.70 (Madsen 1985), 0.75 (CVB 1991), whilst an EAAP Ring Test suggested values nearer 0.7 (Oldham pers.comm).

2. *Digestibility of MTP (DMTP), is estimated to be a constant 0.85,* as recommended by ARC (1980;1984), also Madsen (1985) and CVB (1991).

The amount of DMTP in the estimated MCP supply is therefore:

$$\text{DMTP (g/d)} = 0.75 \times 0.85 \times \text{MCP (g/d)} = 0.6375\text{MCP} \qquad (22)$$

Efficiencies of utilisation of Metabolisable Protein

Efficiency of utilisation of an "ideal" amino acid mixture (k_{aai})

AFRC (1992) describes the limiting efficiency of use of an "ideally" balanced amino acid mixture (k_{aai}) as an animal characteristic. In practical ruminant diets, lower values are achieved, and the term Relative Value (RV), was introduced by AFRC (1992) (equivalent to Biological Value (BV) in non-ruminant animals) to correct for these effects. These depend on particular feeding circumstances, and on the amino acid balance in the Digestible Undegraded Protein (DUP), relative to that in the absorbed amino acids of rumen microbial origin, (DMTP). Users of the system can therefore exercise judgement about the values for RV to be applied in different circumstances, depending on data for RV available.

From data on limiting efficiencies of "ideal" amino acid utilisation across a wide range of circumstances, AFRC (1992) estimate that k_{aai} is at least 0.85. They also decided that k_{aai} for the replenishment of basal endogenous losses of N (BEN, maintenance), will in effect be 1.0 under normal feeding circumstances, as this will be an obligatory demand on the available amino acids.

Thus k_{aai} is 1.0 for maintenance, and 0.85 for all protein synthetic functions.

Relative Value of absorbed amino acid mixtures (RV)

RV will vary according to the mixture of amino acids supplied to the tissues and AFRC (1992) adopted the following values:

Growth	RV	=	0.7
Pregnancy	RV	=	1.0
Lactation	RV	=	0.8
Wool	RV	=	0.3

Working values for efficiency of MP utilisation (k_n)

By combining values for k_{aai} (1.0 or 0.85 as appropriate) and RV, the following working values for k_n are suggested:

Maintenance	k_{nm}	=	1.00
Growth	k_{ng}	=	0.59
Pregnancy	k_{nc}	=	0.85
Lactation	k_{nl}	=	0.68
Wool	k_{nw}	=	0.26

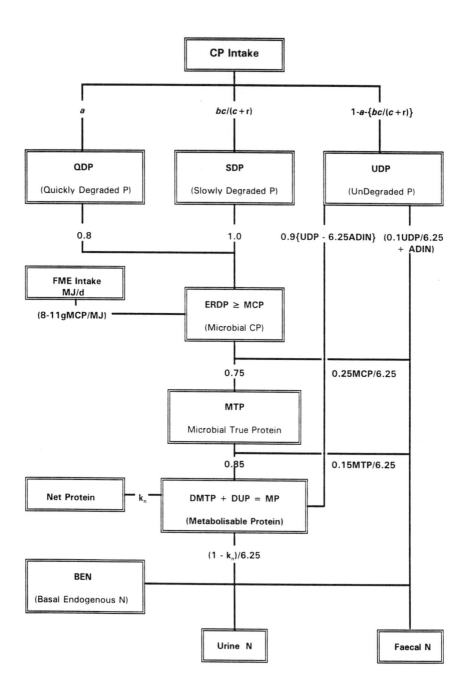

Fig. 1.6: Flow chart of the Metabolisable Protein system.

Chapter Two

Requirements for Metabolisable Energy

Methods of calculating ME requirements

The methods of calculating the ME requirements of cattle and sheep detailed in this Chapter are essentially those recommended in the ARC (1980) Technical Review, but incorporating changes in energy requirements recommended by the advisory services Working Party on energy requirements (AFRC 1990), who tested the ARC (1980) proposals on suitable experimental databases. Apart from a number of changes in activity allowances, they recommended a bias correction factor (C4) when calculating the ME requirements of growing and fattening cattle. The ME requirements of goats are taken (with permission) from the manuscript of the TCORN Working Party on "Nutrition of Goats".

The general equation for the calculation of ME requirements was given in Chapter One:

$$ME \text{ (MJ/d)} = E/k \tag{16}$$

For dairy cattle, lactating sheep and goats, this can be extended to:

$$M_{mp} \text{ (MJ/d)} = C_L\{E_m/k_m + E_l/k_l + E_g/k_g + E_c/k_c\} \tag{17}$$

where C_L is as defined in equation (12) in Chapter One, and the Net Energy values (E) required are defined in the rest of this Chapter.

For growing cattle and sheep, where energy retention (R), is predicted in accordance with the principles outlined in Chapter One, AFRC (1990) give the calculation of ME requirements as:

$$M_{mp} \text{ (MJ/d)} = (E_m/k) \times \ln\{B/(B - R - 1)\} \tag{18}$$

where E_m is the sum of the animal's fasting metabolism (F), and the appropriate activity allowance (A), and B is as defined by equation (14), (values are in Table 1.1), whilst k is defined by equation (15), both given in Chapter One:

$$B = k_m/(k_m - k_f) \qquad (14)$$

$$k = k_m \times \ln(k_m/k_f) \qquad (15)$$

and scaled energy retention (R) is calculated from:

$$E_f \ (MJ/d) = C4(EV_g \times \Delta W) \qquad (37)$$

> *where C4 = 1.15 for bulls and castrated males*
> *= 1.10 for heifers*
> *= 1.00 for growing lambs, since no correction factor was*
> *suggested by AFRC (1990),*

and then: $$R = E_f/E_m \qquad (38)$$

Safety margins

ARC (1980) made no recommendations on the question of safety margins, but MAFF (1976), which recommended the adoption of a simplified ME system, since widely used in practice, included a 5% safety margin in its tables of ME requirements, without giving any statistical argument in support. AFRC (1990) considered the question on a statistical basis, but gave no firm recommendations on the dimensions of a suitable safety margin. Their views can be summarised:

1. *Beef cattle:* Table 5.23, p751, shows that a 5% addition to ME requirement reduces the proportion of cattle underfed by 10%. This is in addition to the inclusion of a bias correction of 1.15 on calculated energy retentions, as on p786. (The example on p787 has the factor of 1.15 in the wrong position - an error confirmed by the authors of TCORN No.5.)
2. *Dairy cattle:* On p758 it is stated that ARC 1980 ME requirements are 10% too low on average. Table 6.11 on p758 shows that a 10% margin reduces the proportion underfed by 30%, whilst a 10% reduction is achieved with a 3% safety margin.
3. *Pregnant ewes:* Table 8.9, p779, shows that a 6% safety margin is needed to reduce underfeeding proportion by 10%.
4. *Growing sheep:* Table 10.9, p785, states that a 15% addition to ME requirements will be needed to achieve the 10% reduction in underfeeding.

No recommendations were made for pregnant cows and lactating sheep, as no suitable databases could be found for testing the ME requirements of ARC (1980). In the light of the above information, and the current practice of using a 5% safety margin on ME requirements in MAFF (1976), *the Sponsors agreed that a 5% safety margin should be added to the ME requirements calculated in accordance with ARC (1980) and AFRC (1990). Accordingly, the Tables of ME requirements in the later Chapters include this agreed safety margin of 5%.*

The Net Energy content of animal products

Calculation of the ME requirements of a ruminant animal as specified above, requires that the Net Energy content (E) of the product, milk, meat, foetus or wool can be estimated, in addition to the Net Energy needed for maintenance purposes. The proposals of ARC (1980) for Net Energy requirements of cattle and sheep are used here, and those of AFRC (1993) for the Net Energy requirements of goats, but they are regrouped by metabolic function and then species and class of ruminant. *No safety margin has been added to the Net Energy requirement equations listed below, since any safety margin is applied to the calculated ME requirements, as above.*

Maintenance requirements

The maintenance ME requirements of cattle, sheep and goats, M_m, are given by:

$$M_m \text{ (MJ/d)} = (F + A)/k_m \tag{39}$$

where F = fasting metabolism and
A = activity allowance as defined below.

Fasting metabolism

Cattle

The fasting metabolism (F) requirements of cattle are given by ARC (1980) as:

$$F \text{ (MJ/d)} = C1\{0.53(W/1.08)^{0.67}\} \tag{40}$$

where C1 = 1.15 for bulls and 1.0 for other cattle.

The factor of 1.08 converts liveweight (W) to fasted body weight as ARC (1980), Appendix 3.II.

Sheep

The fasting metabolism (F) requirements of sheep are given by:

Up to 1 year of age $\qquad F \text{ (MJ/d)} = C1\{0.25(W/1.08)^{0.75}\} \tag{41}$

Over 1 year old $\qquad F \text{ (MJ/d)} = C1\{0.23(W/1.08)^{0.75}\} \tag{42}$

where C1 = 1.15 for entire ram lambs and 1.0 for females and castrates.

Goats

AFRC (1993) give the fasting metabolism (F) of goats as:

$$F \ (MJ/d) = 0.315W^{0.75} \tag{43}$$

No correction from fasted weight to liveweight was made by AFRC (1993) in their recommendation, as other estimates of the maintenance ME requirements of goats agreed with the selected value of $0.315/kgW^{0.75}$.

Activity allowances

ARC (1980) gives the additional energy costs of activity as follows:

Horizontal movement	2.6J/kgW/metre
Vertical movement	28J/kgW/metre
Standing for 24 hours	10kJ/kgW/d
Body position change	260J/kgW

Dairy cattle

AFRC (1990) recommended an increase in the activity allowance (A), for lactating dairy cattle, set at 4.3kJ/kgW by ARC (1980). AFRC (1990) assumed 500 metres walked, 14 hours standing and 9 position changes, giving:

$$A \ (kJ/d) = (1.30 + 5.83 + 2.34)W = 9.47W \tag{44}$$

Activity allowance (MJ/d) for lactating dairy cows is 0.0095W

For pregnant, non-lactating dairy cattle, AFRC (1990) recommend an activity allowance of 0.0071W MJ/d as for housed beef cattle.

Beef cattle

AFRC (1990) recommended an increase in activity allowance above that of ARC (1980). AFRC (1990) assumed horizontal movement of 200 metres, 12 hours standing and 6 position changes:

$$A \ (kJ/d) = (0.52 + 5.00 + 1.56)W = 7.08W \tag{45}$$

Activity allowance (MJ/d) for beef cattle is 0.0071W

Sheep

Housed ewes ARC (1980) did not specify the amount of activity allowance used in calculating the ME requirements of ewes when they were *out-of-doors*, and gave no table for housed ewes. As many pregnant and lactating ewes are now housed, AFRC (1990) assumed that a housed ewe would walk only 50 metres, stand for 14 hours, and make 14 positional changes per day:

$$A \ (kJ/d) = (0.13 + 5.83 + 3.64)W = 9.6W \qquad (46)$$

Activity allowance (MJ/d) for housed, lactating ewes is 0.0096W

Pregnant ewes   AFRC (1990) reduced the activity allowance still further, assuming only 9 hours standing and only 6 positional changes per day:

$$A \ (kJ/d) = (0.13 + 3.75 + 1.56)W = 5.44W \qquad (47)$$

Activity allowance (MJ/d) for housed, pregnant ewes is 0.0054W

Ewes outdoors   ARC (1980) did not specify whether an activity allowance was included in the ME requirements for outdoor ewes in their Table 3.36, and AFRC (1990) assumed it was 10.6kJ/kgW, as for lambs kept outdoors in Table 3.31, in their testing of the ARC (1980) model, but made no recommendation on the activity of ewes outdoors. The ARC (1980) value of 10.6 kJ/kgW might comprise walking 1000 metres horizontally, standing for 12 hours and making 12 positional changes daily. Its activity allowance would then be:

$$A \ (kJ/d) = (2.6 + 5.0 + 3.1)W = 10.7W \qquad (48)$$

Activity allowance (MJ/d) for a lowland ewe out-of-doors is 0.0107W

Hill ewes   Both the distance walked and the vertical movement would be considerably increased for this class of stock, and the AFRC (1993) figures for goats on good quality range are adopted here, ie 5000 metres walking and 100 metres vertical movement per day, in addition to standing for 12 hours and making 12 positional changes, giving:

$$A \ (kJ/d) = (13.0 + 2.8 + 5.0 + 3.1)W = 23.9W \qquad (49)$$

Activity allowance (MJ/d) for ewes on hill grazing is 0.024W

Housed fattening lambs   AFRC (1990) recommended a lower activity allowance for housed fattening lambs than the ARC (1980) figure of 10.6kJ/kgW for lambs *out-of-doors*. The lamb was assumed to walk only 50 metres, stand for 12 hours and make 6 positional changes per day:

$$A \ (kJ/d) = (0.13 + 5.0 + 1.56)W = 6.7W \qquad (50)$$

Activity allowance (MJ/d) for housed fattening lambs is 0.0067W

Goats

Lowland goats   AFRC (1993) accept the ARC (1980) estimates of the components of activity allowances for ruminants as applying to goats. They assume that a goat on lowland pasture will walk 3000 metres horizontally, move vertically 100 metres, stand for 12 hours, and make 12 positional changes daily, giving:

$$A \text{ (kJ/d)} = (7.8 + 2.8 + 5.0 + 3.1)W = 18.7W \qquad (51)$$

Activity allowance (MJ/d) for a lowland goat is 0.019W

Hill and mountain goats   For goats kept on "good quality range", AFRC (1993) increase the distance walked to 5000 metres, and the vertical distance to 100 metres, which with 12 hours standing and 12 positional changes, gives:

$$A \text{ (kJ/d)} = (13 + 2.8 + 5 + 3.1)W = 23.9W \qquad (52)$$

Thus the activity allowance (MJ/d) for goats on hill grazing is 0.024W

Requirements for milk

Cattle

AFRC (1990) recommend that the energy value of milk, $[EV_l]$, can be predicted with adequate precision by using one of the equations of Tyrell & Reid (1965):

$$[EV_l] \text{ (MJ/kg)} = 0.0384[BF] + 0.0223[P] + 0.0199[La] - 0.108 \quad (53)$$

$$[EV_l] \text{ (MJ/kg)} = 0.0376[BF] + 0.0209[P] + 0.948 \qquad (54)$$

$$[EV_l] \text{ (MJ/kg)} = 0.0406[BF] + 1.509 \qquad (55)$$

where [BF] is butterfat, [P] is Crude Protein and [La] is lactose content, g/kg.

The standard errors of estimate of the equations were $\pm$ 0.035, 0.066 and 0.089. The ME required for lactation, M_l, is then calculated from:

$$M_l \text{ (MJ/d)} = (Y \times [EV_l])/k_l \qquad (56)$$

where Y is the milk yield in kg/d.

Sheep

AFRC (1990) recommends the use of the equation of Brett *et al.* (1972) for the prediction of the energy value, $[EV_l]$, of ewes' milk:

$$[EV_l] \ (MJ/kg) = 0.0328[BF] + 0.0025d + 2.2033 \qquad (57)$$

where d is the number of days of lactation of the ewe.

Failing information on the butterfat content, [BF], of ewe's milk, a weighted mean value of 70g/kg butterfat can be used, giving a range of values for $[EV_l]$ of 4.5 rising to 4.7MJ/kg milk over the ewe's lactation.

Šebek & Everts (1992), working with meat producing breeds of sheep, derived an equation for the $[EV_l]$ of such ewes' milk based on the fat, protein and lactose content determined by NIR machine calibrated on cow's milk:

$$[EV_l] \ (MJ/kg) = 0.04194[BF] + 0.01585[P] + 0.2141[La] \qquad (58)$$

The adjusted r^2 value was 0.99 and the residual standard deviation $\pm$ 0.09MJ.

Goats

The Report of the TCORN Working Party on the "Nutrition of Goats" (AFRC 1993) gives the energy content of milk, $[EV_l]$, from two breeds of goats, the Anglo-Nubian and the Saanen/Toggenburg as 3.355 and 2.835MJ/kg milk respectively. The equations of Tyrell & Reid (1965) can be used if the butterfat, protein or lactose contents of the milk are known, as for dairy cattle. The M_l requirement for lactation of these breeds is therefore:

Anglo-Nubian $\qquad M_l \ (MJ/d) = (Y \times 3.355)/k_l \qquad (59)$

Saanen/Toggenburg $\qquad M_l \ (MJ/d) = (Y \times 2.835)/k_l \qquad (60)$

Requirements for growth

Cattle

ARC (1980), p148, gives a quadratic equation to predict the energy value, $[EV_g]$, of weight gains of cattle, for *castrates of medium-sized breeds* as follows:

$$[EV_g] \ (MJ/kg) = \frac{C2(4.1 + 0.0332W - 0.000009W^2)}{(1 - C3 \times 0.1475\Delta W)} \qquad (61)$$

where C3 = 1 when plane of nutrition, L, > 1 and = 0 when L < 1,
C2 corrects for mature body size and sex of the animal, in accordance with the values given in Table 2.1.

AFRC (1990), Table 5.17, suggested a breed classification into early, medium and late maturing types as in Table 2.2.

Table 2.1: Values of correction factor C2 for [EV$_g$] content of liveweight gains in cattle by maturity group and sex.

Maturity type	Bulls	Castrates	Heifers
Early	1.00	1.15	1.30
Medium	0.85	1.00	1.15
Late	0.70	0.85	1.00

Table 2.2: Classification of cattle breeds into maturity groups.

Early	Medium	Late
Aberdeen Angus	Hereford	Charolais
North Devon	Lincoln Red	Friesian*
(Friesian)*	Sussex	Limousin
		Simmental
		South Devon

Later research suggests the Friesian breed should be re-classified as early maturing, see Chapter Six, p77.

The energy retained in the animal's body per day (E$_g$) is then given by:

$$E_g \ (MJ/d) = (\Delta W \times [EV_g]) \tag{62}$$

Sheep

ARC (1980), p106, gives three equations to predict the energy value [EV$_g$], of the liveweight gain of growing lambs:

Non-Merino males $\qquad$ $[EV_g] \ (MJ/kg) = 2.5 + 0.35W$ $\qquad\qquad$ (63)

Castrates $\qquad\qquad$ $[EV_g] \ (MJ/kg) = 4.4 + 0.32W$ $\qquad\qquad$ (64)

Females $\qquad\qquad$ $[EV_g] \ (MJ/kg) = 2.1 + 0.45W$ $\qquad\qquad$ (65)

AFRC (1990) comment that the sex correction for female sheep appears to be inadequate, and suggests that breed differences need to be taken into account. The energy retained per day (E$_g$) is then given by equation (62).

Goats

AFRC (1993) gives a function fitted to the total body energy content (E_g) of castrate male kids in relation to their liveweight:

$$E_g \text{ (MJ)} = 4.972W + 0.1637W^2 \tag{66}$$

which can be differentiated to give the energy content of liveweight gain:

$$[EV_g] \text{ (MJ/kg)} = 4.972 + 0.3274W \tag{67}$$

The E_g requirement is then given by equation (62).

Growth of the fleece in sheep

To the above requirement for growth should be added the energy retention in the growing fleece, which is specified by ARC (1980), Table 1.30, to be 130kJ/d (0.13MJ/d), for a fleece growing at mean growth rate of 5.5g/d, implying an $[EV_g]$ of 23.7MJ/kg. Because of the small amount involved, it is usually ignored. For wool producing sheep, the amount of energy retained in the fleece increases to 0.25MJ/d according to ARC (1980).

Fibre growth in goats

In the case of fibre producing goats, the daily energy retention, E_w, in the fibre for two breeds has been estimated by AFRC (1993) as:

Cashmere goats $\qquad\qquad E_w = 0.08\text{MJ/d} \tag{68}$

Angora goats $\qquad\qquad E_w = 0.25\text{MJ/d} \tag{69}$

Requirements for pregnancy

Cattle

ARC (1980), Table 1.20, gives the total energy retention at time t (E_t, MJ), in the gravid foetus in cattle, assuming a 40kg calf birthweight, as:

$$\log_{10}(E_t) = 151.665 - 151.64e^{-0.0000576t} \tag{70}$$

where t is days from conception.

The daily energy retention (E_c) can then be calculated as AFRC (1990) from:

$$E_c \text{ (MJ/d)} = 0.025W_c(E_t \times 0.0201e^{-0.0000576t}) \qquad (71)$$

where E_t in MJ is calculated as in equation (70)
and W_c is the calf birthweight in kg.

Calf birthweights can be calculated from the equation of Roy (1980):

$$W_c \text{ (kg)} = (W_m^{0.73} - 28.89)/2.064 \qquad (72)$$

where W_m is the mature bodyweight of the dam.

Typical calf birthweight (W_c) values in kg are given in Table 2.3.

Table 2.3: Calf birthweight (kg) as affected by dam's breed.

Breed	Birthweight	Breed	Birthweight
Jersey	26	Sussex	37
Angus	27	Limousin	39
Guernsey	33	Friesian	39
Beef Shorthorn	33	South Devon	43
Ayrshire	35	Simmental	44
Devon	35	Charolais	44
Hereford	36	Holstein	45
Lincoln Red	37		

Sheep

ARC (1980), p8, gives the total energy content at time t (E_t, MJ), for the gravid foetus in pregnant sheep for a 4kg lamb as:

$$\log_{10}(E_t) = 3.322 - 4.979e^{-0.00643t} \qquad (73)$$

AFRC (1990) then calculate the daily energy retention, E_c, as follows:

$$E_c, \text{ MJ/d} = 0.25W_o(E_t \times 0.07372e^{-0.00643t}) \qquad (74)$$

where t is number of days from conception,
and W_o is the total weight of lambs at birth in kg.

Total lamb birthweights for ewes of different bodyweight are given in Table 2.4, using the equations of Donald & Russell (1970), adopted by AFRC (1990).

Table 2.4: Total lamb birthweights (kg) by ewe liveweight (kg) and litter size.

Ewe weight (kg)	Total lamb weight (W_o) (kg)		
	Single	Twin	Triplet
40	3.3	5.4	6.3
50	3.9	6.4	7.5
60	4.5	7.3	8.7
70	5.0	8.2	9.7
80	5.5	9.0	10.8
90	6.0	9.8	11.8

Goats

AFRC (1993) calculated the daily deposition of energy in the gravid foetus of dairy and fibre goats carrying twins or triplets, using equations derived for sheep (Robinson *et al.* 1977). The mean weights of the kids at birth were taken as 3.95kg each for twins and 3.65kg each for triplets of dairy goats, and 2.75kg each for twins and 2.25kg each for triplets of Cashmere goats. The ARC (1980) equations for ewes, (73) and (74) above, are used in calculating the ME requirements of pregnant goats, as they give similar results to those of Robinson *et al.* (1977), using total weight of kids instead of total weight of lambs.

Allowances for liveweight change in lactating ruminants

Cows

ARC (1980), p38, adopted a value of 26MJ/kg *empty-body* weight gain for lactating cattle, equivalent to 26/1.09 = 23.85MJ/kg liveweight gain (ARC 1980, p42), but this was the same as that adopted for adult sheep, because of the wide variation at that time in published estimates for dairy cattle energy values. Recent work based on the serial slaughter and carcass analysis of lactating Holstein/Friesian dairy cows has now been published by Gibb *et al.* (1992), who report mean net energy values [EV_g] of 17.3MJ/kg liveweight loss and 20.9MJ/kg liveweight gain, with an overall mean of 19.3MJ/kg liveweight change. A value of 19MJ/kg liveweight change has been adopted here:

$$[EV_g] \text{ for liveweight change in lactating cows} = 19MJ/kg \qquad (75)$$

For liveweight loss in cows, ARC (1980), p94, specifies that mobilised body reserves can be utilised with an efficiency (k_t) of 0.84 for the synthesis of milk. Thus:

ME from liveweight loss in lactating cows = *(19 x 0.84)/k_l MJ/kg*   (76)

For a dietary energy concentration (M/D) of 11.5MJ/kgDM (q_m = 0.61), k_l is 0.634, so equation (76) gives a value of 25.2MJ of ME per kg liveweight loss.

Ewes

ARC (1980), p24, recommend that the energy content of *empty-body* weight change for sheep should be taken as 26MJ/kg, so correcting by 1.09 to convert to liveweight gain or loss:

[EV$_g$] for liveweight gain in lactating ewes   = *23.85MJ/kg*   (77)

ME from liveweight loss in lactating ewes = *(23.85 x 0.84)/k_l MJ/kg*   (78)

For a dietary energy concentration (M/D) of 11.5MJ/kgDM (q_m = 0.61), k_l is 0.634, so equation (78) gives a value of 31.6MJ of ME per kg liveweight loss.

Goats

AFRC (1994), having reviewed the published data, decided to adopt the ARC (1980) value of 23.9MJ/kg liveweight change for the EV$_g$ of lactating ewes for lactating dairy goats. Equations (77) and (78) for lactating ewes are therefore also to be used for lactating dairy goats. A nominal liveweight loss of 1 kg/week for the first month of lactation, as suggested by INRA (1988), is also adopted.

ME requirement of lactating goats are reduced by 4.6MJ/d in the first 4 weeks.

Chapter Three

Requirements for Metabolisable Protein

Method of calculating MP requirements

Due to the factorial nature of the AFRC (1992) system for the calculation of the total MP requirement of ruminants, the requirement for each relevant metabolic function is calculated separately and then these are summed. These total MP requirements are independent of dietary energy and protein concentrations and plane of nutrition, which affect MP supply, not requirement. *The proposals of ARC (1980) concerning the Tissue Protein, [TP], ie Net Protein, [NP], contents of animal tissues and secretions are relied upon, with only minor exceptions.* The calculation of the Metabolisable Protein Requirements (MPR) is given by:

$$\text{MPR (g/d)} = NP_b/k_{nb} + NP_d/k_{nd} + NP_l/k_{nl} + NP_c/k_{nc} + NP_f/k_{nf}$$

$$+ NP_g/k_{ng} + NP_w/k_{nw} \tag{79}$$

if $NP_g > 0$, and where:

$NP_b = 6.25$ x BEN (g/d) where BEN $= 0.35W^{0.75}$ (g/d) (as ARC 1984)

$NP_d = 6.25$ x $0.018W^{0.75}$ (g/d) (as ARC 1980)

$NP_l = $ Milk yield (kg/d) x milk **true** protein content (g/kg)

$NP_f = \Delta W$ (kg/d) x protein in gain (g/kg) (as ARC 1980)

$NP_g = \Delta W$ (kg/d) x protein in liveweight change (g/kg) of lactating ruminants (as AFRC 1992)

$NP_c = $ protein gain in foetus and gravid uterus (g/d) (as ARC 1980)

$NP_w = 0.8$ x wool growth (g/d) (as ARC 1980)

Note: When there is liveweight loss, and NP_g is negative, then the term NP_g/k_{ng}
becomes NP_g, since the efficiency of mobilisation is assumed in the system
to be 1.0, so that in this case, $NP_g = MP_g$, retaining the negative sign.

Safety margin

An agreed safety margin of 5% has been used in calculating all the summated
Metabolisable Protein requirements tabulated in later Chapters, **but the
calculations given below are without this additional 5%.**

Maintenance requirements

Cattle and goats

The maintenance NP requirements of cattle and goats, NP_m, are the sum of their
Basal Endogenous Nitrogen (BEN or NP_b) needs plus dermal losses as scurf and
hair (NP_d):

$$NP_m \text{ (g/d)} = NP_b + NP_d \tag{80}$$

As $k_{nm} = 1.0$, equation (80) can be converted to MP_m requirements:

$$MP_m \text{ (g/d)} = MP_b + MP_d = 2.30W^{0.75} \tag{81}$$

where MP_b and MP_d are defined by equations (85) and (86) below.

Ewes

Wool growth in ewes is regarded as part of their maintenance requirement for
MP (see equation 89), so that:

$$MP_m \text{ (g/d)} = MP_b + MP_w = 2.1875W^{0.75} + 20.4 \tag{82}$$

Growing lambs

ARC (1980) suggest that wool growth in lambs is proportional to their rate of
liveweight gain, and no allowance for scurf and hair losses is made. Therefore
the maintenance MP for lambs is:

$$MP_m \text{ (g/d)} = 2.1875W^{0.75} \tag{83}$$

Basal Endogenous Nitrogen

The Basal Endogenous Nitrogen (BEN) requirements of cattle, sheep and goats
are the renamed Total Endogenous N (TEN) recommendations of ARC (1984):

$$BEN \text{ (gN/d)} = 0.35W^{0.75} \tag{84}$$

Converting to Crude Protein by the factor of 6.25, and allowing for $k_{nb} = 1.00$, requirement for basal maintenance (MP_b) is:

$$MP_b \text{ (g/d)} = 6.25 \text{ x } 0.35W^{0.75}/1.00 = 2.1875W^{0.75} \qquad (85)$$

Dermal losses as scurf and hair

Cattle and goats

An allowance for dermal losses of protein as scurf and hair (MP_d) should be included, where $k_{nd} = 1.00$:

$$MP_d \text{ (g/d)} = 6.25 \text{ x } 0.018W^{0.75}/1.00 = 0.1125W^{0.75} \qquad (86)$$

Requirements for milk

The efficiency of utilisation of absorbed amino acids for milk protein synthesis (k_{nl}) has been specified as 0.68, so that:

$$MP_l \text{ (g/kg milk)} = \text{(True protein content of milk)}/0.68$$

$$\text{or } 1.471 \text{ x (True protein content of milk)} \qquad (87)$$

Cattle

The *Crude* Protein content of milk is reported by MMB laboratories as percent per litre of milk (P%), which contains about 0.95 *true* protein. The mean density of milk is 1.03kg per litre, so that for a milk *crude* protein of P% per litre:

$$MP_l \text{ (g/kg milk)} = (1.471 \text{ x P\% x } 10 \text{ x } 0.95)/1.03 = 13.57P\% \qquad (88)$$

Sheep

ARC (1980), p47, recommends a value of 7.66g *true* protein N/kg of ewe's milk, equivalent to 7.66 x 6.38 = 48.9g *true* protein/kg milk, which gives:

$$MP_l \text{ (g/kg milk)} = 1.471 \text{ x } 48.9 = 71.9 \qquad (89)$$

Goats

ARC (1980) did not deal with the nutrient requirements of goats, but INRA (1988), p176, quotes the true protein content of goats' milk as 29g/kg. The Report of the TCORN Working Party on the "Nutrition of Goats" (AFRC 1993) gives the *Crude* Protein content of milk from two breeds of goats, the Anglo-Nubian and the Saanen/Toggenburg, as 36 and 29g/kg respectively. Morant (*pers.comm*) found that the *true* protein fraction of goats' milk averaged 0.9 of the *Crude* Protein content, so that the MP_l requirement is therefore:

Anglo-Nubian MP_l (g/kg milk) = 1.471 x 36 x 0.9 = 47.7   (90)

Saanen/Toggenburg MP_l (g/kg milk) = 1.471 x 29 x 0.9 = 38.4   (91)

Requirements for growth

Cattle

ARC (1980), p148, gives a quadratic equation to predict the Net Protein in the weight gains of cattle (NP_f) *for castrates of medium-sized breeds:*

Net Protein in weight gain, NP_f (g/d) =

$$\Delta W\{168.07 - 0.16869W + 0.0001633W^2 \} \times \{1.12 - 0.1223\Delta W\}   (92)$$

where ΔW is as kg/d.

ARC (1980), Table 1.22, gives corrections to the estimates from this equation. They are increased by 10% for bulls and a further 10% for large (late maturing) breeds. Values are reduced by 10% for heifers and a further 10% for small (early maturing) breeds, as in Table 3.1. The relevant efficiency term (k_{nf}) is 0.59, so multiplying the above equation by its reciprocal, 1.695, and inserting the correction factor C6, AFRC (1990):

$$MP_f \text{ (g/d)} = C6\{168.07 - 0.16869W + 0.0001633W^2 \}$$

$$\times \{1.12 - 0.1223\Delta W\} \times 1.695\Delta W   (93)$$

Table 3.1: Values of correction factor C6 for NP_f content of liveweight gains of cattle by maturity group and sex.

Maturity type	Bulls	Castrates	Heifers
Early	1.00	0.90	0.80
Medium	1.10	1.00	0.90
Late	1.20	1.10	1.00

Note: AFRC (1990), Table 5.17, suggested a breed classification into early, medium and late maturing types, given in Table 2.2 in Chapter Two.

Sheep

ARC (1980), p149, gives two equations to predict the protein retention in fleece free liveweight gain (NP$_f$), namely:

Males, castrates $\qquad$ NP$_f$ (g/d) = ΔW(160.4 - 1.22W + 0.0105W^2) $\qquad$ (94)

Females $\qquad$ NP$_f$ (g/d) = ΔW(156.1 - 1.94W + 0.0173W^2) $\qquad$ (95)

where ΔW is as kg/d.

Since MP is used with an efficiency of 0.59 for growth, equations (94) and (95) should be multiplied by 1/0.59 = 1.695 to give:

Males, castrates $\qquad$ MP$_f$ (g/d) = 1.695ΔW(160.4 - 1.22W + 0.0105W^2) $\quad$ (96)

Females $\qquad$ MP$_f$ (g/d) = 1.695ΔW(156.1 - 1.94W + 0.0173W^2) $\quad$ (97)

To the above requirement for growth should be added the MP requirement for protein retention in the growing fleece (MP$_w$), which is specified in equation (104) below. Combining equations (104), (96) and (97) and multiplying out:

Males and castrates

$$MP_f + MP_w \text{ (g/d)} = \Delta W(334 - 2.54W + 0.022W^2) + 11.5 \qquad (98)$$

Females

$$MP_f + MP_w \text{ (g/d)} = \Delta W(325 - 4.03W + 0.036W^2) + 11.5 \qquad (99)$$

Goats

AFRC (1993) give a function fitted to the total Net Protein content (NP$_f$), of castrate male kids in relation to their liveweight:

$$NP_f \text{ (g)} = 157.22W - 0.347W^2 \qquad (100)$$

which can be differentiated to give the protein content of liveweight gains [NPf]:

$$[NP_f] \text{ (g/kg}\Delta W) = 157.22 - 0.694W \qquad (101)$$

The MP$_f$ requirement per kg liveweight gain is then 1/0.59 times equation (101):

$$MP_f \text{ (g/kg}\Delta W) = 266 - 1.18W \qquad (102)$$

Growth of fleece in lambs

Fleece growth in growing lambs is proportional to their liveweight gain (ARC 1980, p17), and the MP requirement is therefore included in the MP requirements for liveweight gain. The equation is:

$$NP_w \text{ (g/d)} = 3 + 0.1 \times NP_f \qquad (103)$$

As the efficiency of utilisation of MP for wool synthesis (k_{nw}) is only 0.26, multiplying by $1/0.26 = 3.846$, the requirement of growing lambs for MP_w is given by:

$$MP_w \text{ (g/d)} = 11.54 + 0.3846 \times NP_f \qquad (104)$$

Growth of fleece in ewes

For ewes, the average protein retained in wool is given as 5.3g/d, based on daily wool growth of 6.6g/d and a protein content of 800g/kg, implying a 2.6kg fleece per annum (see ARC 1980, Table 1.30, p50 and p150). The efficiency of utilisation of MP for wool synthesis (k_{nw}), is 0.26, so that:

$$\text{For ewes only, } MP_w \text{ (g/d)} = 5.3/0.26 = 20.4 \qquad (105)$$

Fibre growth in goats

In the case of fibre producing goats, their daily protein retention (NP_w) in the fibre has been estimated to be 3.6g/d for Cashmere goats, and 10.0g/d for Angora goats (AFRC 1993). The same efficiency of utilisation of MP for fibre synthesis has been adopted for goat fibre as for wool, ie 0.26, so the MP_w requirement in g/d is:

Cashmere goats $MP_w \text{ (g/d)} = 3.6/0.26 = 13.6$ $\qquad (106)$

Angora goats $MP_w \text{ (g/d)} = 10.0/0.26 = 38.5$ $\qquad (107)$

Requirements for pregnancy

Cattle

ARC (1980), p149, gives the daily Tissue Protein (= Net Protein) retention (NP_c), for pregnancy in cattle to produce a 40kg calf as:

$$NP_c \text{ (g/d)} = TP_t \times 34.37e^{-0.00262t} \qquad (108)$$

where t is number of days from conception
and TP_t in kg is given in ARC (1980), Table 1.20, p28, as:

$$\log_{10}(TP_t) = 3.707 - 5.698e^{-0.00262t} \tag{109}$$

TP_t values for other weights of calf (W_c) are directly proportional (see Chapter Two, Table 2.3). The efficiency of utilisation of MP for pregnancy (k_{nc}) is 0.85, so the requirement for MP_c is:

$$\text{For cattle, } MP_c \text{ (g/d)} = 0.85 \times 0.025W_c \times TP_t \times 34.37 \times e^{-0.00262t}$$

$$= 1.01W_c(TP_t \times e^{-0.00262t}) \tag{110}$$

Sheep

ARC (1980), p150, gives the daily Tissue Protein (= Net Protein) retention (NP_c) for pregnancy in sheep to produce a single 4kg lamb as:

$$NP_c \text{ (g/d)} = TP_t \times 0.06744e^{-0.00601t} \tag{111}$$

*where t is number of days from conception
and TP$_t$ in g is given in ARC (1980), Table 1.6, p8, as:*

$$\log_{10}(TP_t) = 4.928 - 4.873e^{-0.00601t} \tag{112}$$

TP_t values for other weights of lamb (W_o) are directly proportional (see Chapter Two, Table 2.4). Since the efficiency of utilisation of MP for pregnancy (k_{nc}) is 0.85, the requirement for MP_c is given by:

$$MP_c \text{ (g/d)} = 0.25W_c(0.079TP_t \times e^{-0.00601t}) \tag{113}$$

Goats

AFRC (1993) calculated the daily deposition of protein in the gravid foetus of dairy and fibre goats carrying twins or triplets, using equations derived for sheep (Robinson *et al.* 1977) rather than equation (111) above, derived from the ARC (1980) equations for ewes. The differences are quite minor, so the use of equation (111) is recommended here for single and twin foetuses. The mean weights (W_o) of the kids at birth were taken as 3.95kg each for twins and 3.65kg each for triplets of dairy goats, and 2.75kg each for twins and 2.25kg each for triplets of Cashmere goats.

Allowances for liveweight change in lactating ruminants

Cows

ARC (1980), p38, gives the Net Protein content of *empty-body* weight change, [NP_g], as 150g/kg, equivalent to $150/1.09 = 138$g/kg *live*weight gain (ARC

1980, p42). Gibb *et al.* (1992) found a mean net change in the body composition of lactating dairy cows of 126gCP in 0.8kg/d liveweight loss, equivalent to 158g/kg liveweight loss, but gave no estimate for the Net Protein content of liveweight gains. The differences in estimates will have only a small effect upon the calculated MP requirements, so the ARC (1980) figure of 138g/kg liveweight change has been used here. Using the appropriate efficiency factor ($k_{ng} = 0.59$):

$$MP_g \text{ (g/kg) liveweight gain in cows is } 138/0.59 = 233 \qquad (114)$$

For liveweight loss in cows, ARC (1980), p149, implies an efficiency of utilisation of mobilised body protein of 1.0, so that:

Mobilised Net Protein = Metabolisable Protein

The mobilised amino acids are then utilised with the same efficiency for protein synthesis as absorbed microbial and feed amino acids. The Metabolisable Protein (MP_g) contributed from liveweight loss will therefore be 138g/kg. Thus:

$$MP_g \text{ (g/kg) liveweight loss in cows } = 138 \qquad (115)$$

Ewes

ARC (1980), p24, gives the Net Protein content of liveweight loss in lactating ewes [NP_g] as 130g/kg *empty-body* weight. Here also, Net Protein = Metabolisable Protein. As 130g/kg *empty-body* weight is equivalent to 119g/kg *live*weight, then:

$$MP_g \text{ (g/kg) liveweight loss in ewes } = 119 \qquad (116)$$

The Tissue (Net) Protein content in *empty-body* gain in ewes is stated on p24 of ARC (1980) to be only 90g/kg, equivalent to 83g/kg *live*weight gain. The MP_g for liveweight gain in ewes will be given by $83/0.59 = 140$, so:

$$MP_g \text{ (g/kg) liveweight gain in ewes } = 140 \qquad (117)$$

Goats

INRA (1988) suggest that the PDI (= Metabolisable Protein) deficit of dairy goats can be 85-25g/d in the first two weeks of lactation, associated with a liveweight loss of 1kg per week. AFRC (1994) suggest that MP allowances for lactating goats may be reduced by 30gMP/d, a total of 900gMP in the first month of lactation, which for the nominal 1.0kg liveweight loss per week allowed for energy, implies an NP_g for liveweight loss of 143g/kg, comparable to the ARC (1980) recommendation of 119g/kg for ewes and 138g/kg for cows.

MP requirements of lactating goats may be reduced by 30g/d in the first month of lactation.

Chapter Four

Feed Evaluation and Diet Formulation

Nutrient values for ruminant feeds

Diet formulation requires two sets of matching information, the nutritive requirements defined in the selected scientific units for the species and class of animal under consideration, and equivalent information for each of the feeds selected for inclusion in the diet to be formulated. Preceding chapters have dealt at length with the animals' requirements for energy as MJ of ME per day, and protein requirements as g MP per day. This chapter will deal with the estimation and prediction of the matching energy and protein values for feeds, and their assembly into diets that meet the animals' requirements. In the case of ME, then ME as MJ/kgDM is the required parameter, but in the case of protein, the two parameters ERDP and DUP, both as g/kgDM, are needed, since MP supply is only calculated from them for the assembled diet.

Metabolisable Energy values of ruminant feeds

MAFF (1976), continued in MAFF (1984), gave a number of prediction equations for the ME values of various classes of feeds, most of which have now been superseded, except those based on the *in vitro* digestibility technique of Tilley & Terry (1963). They have been replaced either by cellulase enzyme techniques (Dowman & Collins 1982), or else by Near-Infrared Reflectance Spectroscopy (NIR) methods in widespread use in the industry, based on reference samples of known *in vivo* digestibility and [ME] value. Barber *et al.* (1984) summarised the situation as at that date, but Givens *et al.* (1988;1989;1990;1992) and ADAS (1994a) have published a number of papers since which replace and update those of Barber *et al.* (1984). A selection of these published equations is given below. In the case of grass hays, conversion from [DOMD] to [DE] used the published mean values for these parameters in Moss & Givens (1990). The reader is referred to the full publications or to the ADAS Feed Evaluation Unit, Drayton Manor Drive, Stratford-upon-Avon for fuller details, and for reference samples for NIR calibration.

Analytical methods used

MADF Modified Acid Detergent Fibre   Clancy & Wilson (1966)
NCD   Neutral detergent cellulase [DOMD]   Dowman & Collins (1982)
NCDG Neutral detergent cellulase + gammanase [DOMD]   MAFF (1993)
IVD   *in vitro* digestibility (DOMD)   Tilley & Terry (1963)

			r^2	
Fresh grass[1]	ME (MJ/kgDM)	= 16.20 - 0.0185[MADF]	0.59	(118)
		= 3.24 + 0.0111[NCD]	0.68	(119)
		= -0.46 + 0.0170[IVD]	0.61	(120)
Grass hays[2]	ME (MJ/kgDM)	= 15.86 - 0.0189[MADF]	0.67	(121)
field cured	ME (MJ/kgDM)	= 4.28 + 0.0087[NCD]	0.49	(122)
		= 2.67 + 0.0110[IVD]	0.83	(123)
barn dried	ME (MJ/kgDM)	= 1.80 + 0.0132[NCD]	0.86	(124)
		= 0.61 + 0.0148[IVD]	0.90	(125)
Dried grass[3] (a)	ME (MJ/kgDM)	= 16.90 - 0.0224[MADF]	0.61	(126)
(high temperature)		= -0.59 + 0.0154[NCD]	0.81	(127)
		= -1.82 + 0.0195[IVD]	0.76	(128)
Dried lucerne[4]	ME (MJ/kgDM)	= 13.90 - 0.0164[MADF]	0.58	(129)
(high temperature)		= -0.61 + 0.0151[NCD]	0.79	(130)
		= -0.49 + 0.0163[IVD]	0.61	(131)
Grass silage[5]	ME (MJ/kgDM)	= 15.0 - 0.0140[MADF]	0.14	(132)
(also by NIR)		= 5.45 + 0.0085[NCD]	0.20	(133)
		= 2.91 + 0.0120[IVD]	0.24	(134)
Maize silage[6] (b)	ME (MJ/kgDM)	= 13.38 - 0.0113[MADF]	0.46	(135)
(also by NIR)		= 3.62 + 0.0100[NCD]	0.64	(136)
Cereal straws[7]	ME (MJ/kgDM)	= 0.53 + 0.0142[IVD]	0.36	(137)
" *ammoniated*[7]	ME (MJ/kgDM)	= 2.24 + 0.0098[IVD]	0.56	(138)
" *alkali treated*[8]	ME (MJ/kgDM)	= 1.62 + 0.0121[NCD]	0.35	(139)
(also by NIR)		= 2.54 + 0.0093[IVD]	0.25	(140)
Compound feed[9] (c)	ME (MJ/kgDM)	= 0.0140[NCDG] + 0.025[EE]		(141)

Notes: *All analyses are as g/kg oven dry matter*
   (a)   Research funded by the British Association of Crop Driers and MAFF
   (b)   Research funded by the Maize Growers Association and MAFF
   (c)   Research funded by MAFF/DAFS and UKASTA. RMSE $\pm$0.24MJ

References:

1	Givens *et al.* (1990).
2	Moss & Givens (1990).
3	Givens *et al.* (1992).
4	Givens (1989).
5	Givens *et al.* (1989).
6	Givens (1990).
7	Givens *et al.* (1988).
8	Moss, et al. (1990).
9	Thomas *et al.* (1988), MAFF (1991; 1993)
10	ADAS (1994a)

Prediction of ME values from *in vivo* or *in vitro* [DOMD] values

Accurate measurement of the ME values of feeds or diets requires the execution of *in vivo* digestibility trials with either sheep or cattle, and measurement of the Gross Energy of feed, faeces, urine and methane, although the latter is often estimated using the Blaxter & Clapperton (1965) equation. Understandably, most of such measurements are made in centres specially equipped so to do, but many other research centres carry out *in vivo* digestibility trials, assessing either dry matter or organic matter digestibilities. Others routinely measure *in vitro* digestibility, usually expressed as [DOMD] or [OMD] either as per cent or g/kg.

Because of the relative constancy of both the GE of Digestible Organic Matter for most forages other than silage, and the proportions of energy lost in urine and methane, highly significant correlations are normally found between ME and [DOMD] values, such that approximate conversions of [DOMD] values to ME values may be made. Barber *et al.* (1984) examined their *in vivo* data for ME and [DOMD] values for most types of forages and found that a pooled regression could be fitted:

$$ME \text{ (MJ/kgDM)} = 0.0157[DOMD] \qquad r^2 = 0.83 \qquad (142)$$

where [DOMD] is as g/kg dry matter.

Prediction of the ME of grass silage by NIR spectroscopy

Examination of the best prediction equations (133 and 134) for the [ME] of grass silages in the preceding section reveals that poor correlations are reported, with only one quarter of the variance accounted for. Work by Givens *et al.* (1989) showed that a primary cause was the variation in the [GE] content of the DOM of grass silages, and that improved predictions were obtained if [GE] was introduced as a secondary variable. Such an approach is impracticable in routine laboratories, since the measurement of [GE] is time consuming at present.

The advent of NIR spectroscopy instruments offered the possibility of developing rapid, accurate and routine methods of predicting the [ME] of grass silages and cereal straws from estimates of the digestibility of the organic matter of the forage. This has been confirmed by the work of Barnes *et al.* (1989) and Barber *et al.* (1990), leading to a recent industry-wide agreement on standard procedures for the use of NIR for this purpose (ADAS/DANI/SAC/UKASTA 1993). The approach used is a two-stage one as follows:

1. Prediction of the Organic Matter Digestibility [OMD] as g/kg oven DM using NIR instruments calibrated with samples of known *in vivo* [OMD].
2. Conversion of the silage predicted [OMD] value to [ME] as MJ/kg on a Corrected DM basis [CDM].

Prediction of [OMD] of grass silage

The reason for the choice of [OMD] instead of the widely used [DOMD], is that NIR instruments cannot adequately measure the ash content of forages, so that this source of variation needs to be eliminated. Accurate predictions of silage [OMD] have been obtained by calibrating NIR instruments on 122 samples of grass silage of known *in vivo* [OMD] and [ME] content, with 65% of estimated [ME] values being within ± 0.3MJ of the *in vivo* measurements (Barber *et al.* 1990). Details of the NIR calibrations can be leased from ADAS, SAC or DANI, both for scanning and 19 filter NIR instruments.

Conversion of [OMD] to [ME]

The volatile compounds present in fresh silage (primarily fatty acids and ethanol) have high [GE] values, so that correcting for their energy value is an important part of the estimation of the energy consumed by the animal, but there may be some loss of these to the atmosphere before consumption by the animal. This is dealt with by the use of the term Corrected Dry Matter [CDM], defined below, and expressing silage [ME] values on this base, rather than Oven Dry Matter [ODM]. There are four steps in the conversion process:

1. $\qquad$ $[DOMD_o]$ (g/kg) = [OMD] x (1000 - total ash)/1000 $\qquad$ (143)

where $[DOMD_o]$ is the Digestible Organic Matter as g/kg of Oven Dry Matter.

This equation allows for the diluting effect of total ash (g/kgODM) on digestible energy value. Total ash is measured by incineration overnight at 450°C.

2. $\qquad$ $[DOMD_c]$ (g/kg) = 1000 - {(1000 - $[DOMD_o]$) x [ODM]/[CDM]} (144)

where $[DOMD_c]$ is the Digestible Organic Matter including volatiles (g/kgCDM).

Equation (144) allows for the volatile compounds' contribution to the digested organic matter, DOM. Corrected Dry Matter [CDM], is estimated as follows:

3. $$[CDM] \ (g/kg) = 0.99[ODM] + 18.2 \tag{145}$$

Better methods of correcting for the volatile compounds in silages are under development. The last step in the prediction is to convert the $[DOMD_c]$ value to [ME] as MJ/kgCDM:

4. $$[ME] \ (MJ/kgCDM) = 0.016[DOMD_c] \tag{146}$$

The mean literature value for the ratio of [ME] to [DOMD] is 0.168, where methane is measured, appreciably higher than the general mean for all forages found by Barber *et al.* (1984), equation (134) above. The lower value of 0.16 adopted here includes a 5% safety allowance for possible losses of volatiles during feeding out.

[ERDP] and [DUP] values of ruminant feeds

Chapter One gave the theoretical derivation of the units [ERDP] and [DUP], and the effects of the level of feeding (L) in calculating the values to be used in a diet at the selected feeding level. The feed parameters from which these feeding values are derived are the Crude Protein content of the feed [CP], and the fitted parameters from the Ørskov & McDonald (1979) equation, *a, b* and *c*, using data from measurements of N disappearance from dacron bags suspended in the rumen of either cattle or sheep. At the time of writing this Manual, the UK public domain database for the *a, b* and *c* values is rather limited, so that some approximations may be needed to assign temporary working values to feeds included in the Tables of Feed Composition in Appendix I.

Conversion of existing [RDP] and [UDP] values to [ERDP] and [DUP]

There are published data available on protein (N) degradability (dg), but these are often single estimates at a particular retention time, commonly 12 hours, equivalent to a fractional outflow rate of about 0.05. Even if multiple estimates were made of N loss from the dacron bag at intervals, and the Ørskov & McDonald (1979) equation was used to derive an effective degradability (p) the values of the fitted parameters *a, b* and *c* have often remained in laboratory notebooks. If it is assumed that existing dg values can be taken as valid for an outflow rate of 0.05 (dg_5), then estimates of [RDP] and [UDP] for r = 0.05 are available, and can be converted to [ERDP] and [DUP] values as shown below.

Estimation of [QDP], the "a" fraction

The Ørskov & McDonald (1979) zero time intercept *a*, is determined by washing the specially prepared feed sample placed within the standard dacron bag with cold tap water, either by hand or in a modified washing machine, as recommended by AFRC (1992), Appendix II. No access to animals is required in order to carry out this measurement and it is unaffected by outflow rate. [QDP] is particularly important in ensiled feeds, because it may be more than 0.5 of the [RDP] fraction, and it is used with the lower efficiency of 0.8. Hot water soluble N has been used to characterise grass silages, and various buffer solutions have been used as extractants of N from feeds. The proportions of N extracted by a buffer solution for many raw materials are listed in the Nordic AAT/PBV system (Madsen 1985; Hvelplund & Madsen 1990). For the limited number (13), of concentrate feeds for which there is an a value in the MAFF (1990) Tables and a matching buffer solubility in Hvelplund & Madsen's (1990) Feedstuffs Table, a significant regression was obtained:

$$a = 0.06 + 0.61 \text{ x buffer solubility } (r = 0.6) \qquad (147)$$

The method used used by Madsen (1985) was to take 500mg of dried feed ground through a 0.7mm screen, place in a 100ml centrifuge tube, add 60ml buffer solution and incubate @ 38°C for 1 hour. After centrifugation, the washing procedure was the same as the zero time dacron bag samples. The buffer had the following composition, all as g/l: Na_2HPO_4, 4.6; $NaHCO_3$, 9.8; KCl, 0.6; NaCl, 0.5; $MgCl_2$, 0.06; $CaCl_2$ 0.04.

This method of estimating a values has been used to fill out the Tables of Feed Composition in Appendix I of this Manual. Values so estimated are annotated. Work is in progress to use cold water extraction of feed suspended in coarse filter paper as a rapid method of estimating the [QDP] fractions of feeds (Cockburn et al. 1993). Alternatively, an estimate of the proportion of the [RDP] that is quickly degraded may be made, using values from other comparable feeds for guidance. Then [QDP] = a x [CP].

Estimation of [SDP]

The amount of [SDP] is [RDP] - [QDP], both of which are now to hand for a fractional outflow rate of 0.05. When fitting the Ørskov & McDonald (1979) function to degradability data, it is usual to constrain $a + b$ to be less than 1, but the fraction of feeds that are completely undegradable (u), when the time (t) is infinite, ie the asymptotes in Figs 1.2 and 1.3 in Chapter One, are fairly small, as summing the a and b fractions in the Tables of Feed Composition in Appendix I shows. The mean value for u for concentrate feeds is 0.05, with rather larger values of about 0.20 for forages. Since:

$$a + b + u = 1 \qquad (148)$$

the dimensions of b can be estimated if a and u have been estimated as suggested. However, to estimate [SDP] values at other outflow rates, some estimate of the rate of degradation (c) is required.

Assessment of rate of degradation, "c"

In terms of the parameters b and c, [SDP] at an outflow rate of 0.08 is given by equation (28), using r = 0.08. In order to estimate [SDP] at other outflow rates, the value of c also needs to be estimated. This can be done by rearranging equation (26) and substituting dg_8 values for p as follows:

$$c = r(a - dg_8)/(dg_8 - a - b) \qquad (149)$$

Using r = 0.08 (or 0.06 for dg_6 values) the equation can be solved for c and then equation (28) used to calculate the required [SDP] value, followed by equation (29) to calculate the [ERDP] value. The Nordic feed database values are calculated using an outflow rate of 0.08, so that their data are dg_8 values. The INRA (1988) database values (their Table 13.3), are calculated for an outflow rate of 0.06, ie are dg_6 values, as are those of Van Straalen & Tamminga (1990). The working range of c values appears to be 0.02 (fishmeal) to 0.50 (wheat). Most heat treated feeds have low c values, 0.02-0.04, with forages varying 0.05-0.20, depending on type. Barley (0.3) and wheat (0.5) are the most rapidly degraded common feeds, when fed in the dried, ground form.

Estimation of [DUP]

If dg_8 values are available for a particular feed, then the value of [UDP] is given by [CP] x (1 - dg_8). The estimation of [DUP] then only requires that some estimate of the digestibility of [UDP] be made. INRA (1988), Table 13.3, gives mean values for the digestibility of UDP in the small intestine (*dsi*) for about 80 common feeds. INRA (1988) feed class means are given in Table 4.1. Alternatively, equation (33) may then be used to estimate [DUP] if a value for [MADF] (or [ADF]) is available, using [ADIN] = [MADF]/62.5 (see also Chapter One, Table 1.3). Equation (33) implies that the digestibility of [UDP] is normally high (0.8-0.9) for feeds high in [UDP], which are also low in fibre, eg many of the oil seed meals and fishmeal, but appreciably lower for forages, where values of nearer 0.5-0.6 have been estimated.

Mean N degradability parameters by class of feed

AFRC (1992) showed that the precision of dacron bag N disappearance measurements was not good, so that undue reliance should not be placed on estimates of [ERDP] and [DUP] obtained by the use of feed class means. However, for many of the concentrate feeds which are traded internationally, the use of data from Denmark, France and The Netherlands should not be a major source of error, other than that due to laboratory variation. Where figures on the same raw material are available from several sources, the degree of agreement has been found to be satisfactory. Feed class mean values for the parameters a, b and c, [ADIN] values (g/kgDM) and mean digestibilities of [UDP], either predicted from [ADIN] or as *dsi* values (INRA 1988) are given in Table 4.1:

Table 4.1: Mean values for N degradability parameters by class of feed.

Feed class	a	b	c	u	[ADIN]*	dsi†
Fresh forages	0.24	0.67	0.12	0.09	1.2	0.75
Roots	0.25	0.65	0.41	0.10	1.2	0.60-0.95
Grass and legume silages	0.59	0.31	0.13	0.11	1.2	0.60
Cereal silages, inc. maize	0.69	0.20	0.10	0.10	2.2	0.70
Grass hays	0.22	0.60	0.08	0.18	1.2	0.70
Legume hays	0.20	0.65	0.29	0.15	2.0	0.75
Cereal straws	0.30	0.50	0.12	0.20	1.0	0.70
Cereals	0.47	0.48	0.27	0.05	0.4	0.95
Legume seeds	0.41	0.57	0.16	0.01	0.5	0.60
Cereal byproducts	0.36	0.55	0.09	0.09	1.0	0.75-0.95
Beet and citrus pulps	0.39	0.57	0.05	0.04	1.2	0.75
Oil meals, high fibre	0.24	0.69	0.11	0.08	3.3	0.85
Oil meals, low fibre	0.14	0.79	0.09	0.07	2.1	0.90

* Expressed as g/kgDM.   *† From INRA (1988), Table 13.3.*

Fermentable Metabolisable Energy

The Fermentable Metabolisable Energy [FME] content of a feed or diet is defined by AFRC (1992) (see Chapter One) as:

$$[FME] \ (MJ/kgDM) = [ME] - [ME_{fat}] - [ME_{ferm}] \qquad (5)$$

The common fermentation acids in silages and fermented products have [ME] (actually [GE] values in this case) as in Table 4.2:

Table 4.2: Gross Energy contents (MJ/kg) or (MJ/g) of fermentation acids.

Fatty acid	Formula	MJ/kg	MJ/g
Acetic acid	(CH_3COOH)	14.6	0.0146
Propionic acid	(C_2H_5COOH)	20.8	0.0208
Butyric acid	(C_3H_7COOH)	24.9	0.0249
Valeric acid	(C_4H_9COOH)	28.0	0.0280
Lactic acid	$(CH_3CHOH.COOH)$	15.2	0.0152

If the amounts of fermentation acids are not known, an average value of 0.10 x [ME] of feed should be applied for grass silage. For brewery and distillery byproducts, a correction of 0.05 x [ME] may be used (Webster 1992):

For grass silage: [FME] (MJ/kgDM) = 0.90[ME] - [ME$_{fat}$] (150)

For brewery and distillery byproducts:

[FME] (MJ/kgDM) = 0.95[ME] - [ME$_{fat}$] (151)

As the extent of fermentation in silage is known to be related to the dry matter content of the ensiled crop, a prediction of [FME] can be made using an equation developed by the ADAS Feed Evaluation Unit (ADAS 1991):

$$[FME]\ (MJ/kgDM) = [ME] \times (0.467 + 0.00136[ODM] - 0.00000115[ODM]^2)$$ (152)

where [ODM] is Oven Dry Matter, g/kg, and $r^2 = 0.38$.

[FME] values obtained from equation (152) as a function of the [ME] and [ODM] content of grass silages are given in Table 4.3.

Table 4.3: [FME] content (MJ/kgDM) of grass silage predicted from [ME] (MJ/kgDM) and Oven Dry Matter [ODM] content (g/kg).

[ME] of silage (MJ/kgDM)*	Oven Dry Matter [ODM] (g/kg)								
	175	200	225	250	275	300	350	400	500
9.5	6.4	6.6	6.8	7.0	7.2	7.3	7.6	7.9	8.2
10.0	6.7	6.9	7.2	7.4	7.5	7.7	8.0	8.3	8.6
10.5	7.0	7.3	7.5	7.7	7.9	8.1	8.4	8.7	9.0
11.0	7.4	7.6	7.9	8.1	8.3	8.5	8.8	9.1	9.5
11.5	7.7	8.0	8.2	8.5	8.7	8.9	9.2	9.5	9.9
12.0	8.0	8.3	8.6	8.8	9.1	9.3	9.6	9.9	10.3

* *The use of Corrected Dry Matter, [CDM], (= 0.99[ODM] + 18g/kg) as the base for estimating the [ME] values of silages is in agreed use. Table 4.3 can be used for [ME] values expressed either way, to give [FME] values on the appropriate basis, but ensuring that the dry matter value used has been corrected to [ODM] if need be.*

The average proportion of [FME]/[ME] in grass silage made in clamps (mean [ODM] 252g/kg), was found by ADAS (1991) to be 0.726 ± 0.078, with a range of 0.41 to 0.86, whilst obviously butyric silages ([ODM] 187g/kg), averaged 0.66 ± 0.067, ranging 0.55 to 0.80. Big bale silages, with much higher [ODM] levels (348g/kg), and restricted fermentations, achieved a mean [FME]/[ME] of 0.820 ± 0.074, range 0.67 to 0.91.

The mean [TDM] of 26 samples of maize silage reported in MAFF (1990) was 278g/kg and the mean fermentation acid contents reported were lactic, 68.7; acetic, 32.7; propionic, 2.6; butyric, 3.8; and valeric, 0.6g/kg; giving a value for [ME_{ferm}] of 1.7MJ/kg. However these maize silage samples were from outdated and late maturing varieties, and few recent figures are available for early maturing varieties. Phipps *et al.* (*pers.comm*) determined the fermentation acid content of maize silage with a Toluene Dry Matter content of 354g/kg and found 69.4g/kg lactic acid, but only 12.3g/kg acetic acid and no propionic or butyric acid detectable, to give a lower [ME_{ferm}] of 1.2MJ/kgDM. With a typical [EE] of 29g/kg, this gives an [FME] of 9.0MJ/kgDM, and [FME]/[ME] = 0.80.

Methods of diet formulation

Hand formulated or computer assisted procedures

Diet formulation to meet specified requirements of ME and MP within the animal's dry matter appetite, requires the assembly of a matrix with a total of seven columns, although a column for Crude Protein (CP) may be added if desired. A complete diet formulation will also include columns for carbohydrate fractions, minerals and trace elements, but these lie outside the scope of this Manual. The feed contributions of minerals are all additive, and there are no interactions with dietary characteristics, except in the case of P availability, where AFRC (1991a) have suggested that it varies with type of feed.

Hand steering of diet formulation, albeit computer assisted, is commonplace, and the recommended continuous function (equation 25), relating outflow rate (r) to level of feeding (L), means that the exact [ERDP] and [DUP] values for the currently displayed feeding level can be used in such diet calculations. Similarly, the appropriate MCP yield (y), from equation (34), for the current value of (L) can be used in calculating the total MP supply from the diet. *In the many diet formulation examples that follow in this Manual, feed values are taken at the nearest appropriate feeding level, 1, 2 or 3, so that the reader can refer to the appropriate sets of values for [ERDP] and [DUP] in the Tables of Feed Composition in Appendix 1. Values for microbial yield (y) however, have been calculated for the exact value of L in the example.*

Diet checking procedure

The following three principles should be borne in mind when carrying out calculations on diets formulated using the combined ME and MP systems:

1. *Energy is always the first limiting factor* upon the level of animal production achieved by feeding the formulated diet. Estimates of voluntary feed intake are therefore important as determining ME intake.
2. *The adequacy of the ERDP supply must be considered next,* in order to maximise MCP synthesis by rumen microbes, before considering whether the MP supply meets requirements. It is possible to formulate diets meeting MP requirement, by including high levels of DUP, and inadequate amounts of ERDP to match the FME in the diet but these this would reduce both MCP synthesis and feed intake, restricting energy intake still further.
3. *The adequacy of the MP supply can then be checked,* as can the consequences of surpluses or deficiencies of MP. Inadequate MP supply probably explains the low milk protein levels produced by some dairy cow diets, whilst responses to additional MP at the same ME input, ie extra DUP, have been observed with both dairy and beef cattle (see p66 and p83).

Bearing these principles in mind, the calculations required are as follows:

Example dairy cow diet:

Assumptions:　　600kg cow giving 30kg milk with 4.04% fat and 3.28% protein per litre and losing 0.5kg/d. This gives:
MER = 205MJ, L = 3.3, MPR = 1625g, r = 0.08

Using [ERDP] and [DUP] values for feeds for an outflow rate of 0.08/h:

Feed name	Fresh wt, kg	DM kg	ME MJ	FME MJ	CP g	ERDP g	DUP g	MP g
Grass silage '92	40.0	10.0	110	81	1420	900	180	
Dried beet pulp	2.3	2.0	25	25	206	86	86	
Rolled barley	2.0	1.7	22	21	194	143	31	
Maize gluten feed	2.3	2.0	25	23	414	260	82	
Rapeseed meal	2.3	2.0	24	22	800	530	156	
Totals	**48.9**	**17.7**	**206**	**171**	**3034**	**1919**	**535**	***1745**
Requirements/limits		18.3	205	*173		*1898	*415	**1625**

*Note: The derivation of figures marked * are indicated by * below.*

1. *DM intake* for a concentrate intake of 9kg fresh weight and a milk yield of 30.8kg/d is given as 18.3kg in Table 5.2. This diet is therefore within normal appetite limits.
2. *Diet M/D* is 206/17.7 = **11.6MJ/kgDM**, close to the M/D value of 11.5 used to calculate the dairy cow ME requirements in Table 5.1. Computer software will recalculate the ME requirement as the diet M/D varies.

3. *ME supply* of **206MJ/d** just exceeds the requirement given in Table 5.1

4. *Diet CP concentration* is 3034/18.0 = **171g/kgDM**. This calculation is not essential, but most nutritionists like to see the overall diet [CP].

5. *ERDP/FME ratio* is 1919/171 = **11.22g/MJ FME**. For a feeding level (L) of 3.3, the MCP yield (y) is 11.1g/MJ FME in Table 1.4 derived from equation (34), so that an ERDP/FME ratio of at least 11.1 is required here.

6. *ERDP required* is therefore 171 x 11.1 = ***1898g/d**, indicated as the ERDP requirement * above. As the diet supplies 1919g/d, means that ERDP is not limiting MCP supply or feed intake, but the excess of 21g/d is small.

7. *MCP supply* is calculated from the limiting factor, either ERDP or FME:

From FME   MCP = y x FME = 11.1 x 171 = **1898g/d**
From ERDP   ERDP = MCP = **1919g/d**

The smaller value, 1898g/d, is the amount of MCP predicted to be supplied by this diet, which is marginally limited by FME, with a small surplus of ERDP. The ERDP supply from this diet requires 1919/11.1 = ***173MJ** FME, indicated as a notional requirement above.

8. *MP supply* is calculated from equation (23):

MP (g/d) = 0.6375MCP + DUP = 0.6375 x 1898 + 535 = ***1745**

which exceeds the MP requirement of 1625g/d by 120g/d, probably sufficient to raise the milk protein level above the assumed 3.28% per litre.

It can also be concluded that the amount of DUP supplied by the diet, 535g/d, could be reduced by 120g/d, by altering the choice or amount of feeds included, since the amount of DUP required to increase the DTMP supply from 1898gMCP to meet the MP requirement of 1625g/d is:

DUP requirement (g/d) = 1625 - 0.6375 x 1898 = ***415**

Computer software usually displays these notional requirements for FME, ERDP and DUP as a method of indicating which factor is limiting MCP supply, and whether the amount of DUP supplied is adequate. *Subsequent diet calculation examples will use this method of displaying the diet balances of FME, ERDP and DUP. In practice, ERDP and DUP totals should show a small surplus above notional requirements, whilst the total FME will normally be slightly in deficit.*

The option to use the Variable Net Energy values of feeds (E_{mp}) in diet formulation can also be used for the hand steered computer assisted procedure outlined above, replacing the ME column in the matrix by E_{mp} values calculated for the level of animal production specified (see Chapter One for details). Hand calculation of the required E_{mp} values is not recommended. Provided that *the E_{mp} requirements have the 5% safety margin added* to the results of using the equations in Chapter Two, diets which vary in M/D concentration and total dry matter intake can be formulated successfully.

Computer Linear Programming methods

If it is desired to formulate a diet using Linear Programming techniques, then the use of the appropriate E_{mp} values for the selected feeds is essential (see Chapter One for details). *A safety margin of 5% must be added to the E_{mp} requirements*, if the diets calculated are to agree with those based on the ME requirement Tables in this Manual. The technique will also generate accurate information about the relative feeding values and prices of the feeds offered for consideration. Further details will not be given here, but suitable computer software is available.

Using Dry Matter intake functions

It cannot be stressed too much that diets must be formulated to be within expected dry matter appetite constraints, which is why it is the first check to be made in the sequence listed above. Unrealistic expectations of voluntary silage intakes are the commonest cause of energy deficits in dairy herds.

Most dry matter intake prediction equations recommended by ARC (1980) and AFRC (1990;1991b), whether for forage voluntary intake or total DM intake, include a term for the concentrate portion of the diet (C). The relevant equations are dealt with in detail in Chapters Five to Eight. Thus every addition of a concentrate feed to the basal forage in a hand steered formulation, alters the expected forage intake and the total DM intake. Other equations for beef cattle (ARC 1980) are dependent additionally on the ME concentration of the diet (M/D or q_m), also altered by each addition of concentrate to the diet. If computer assistance is being used, then the alterations in DM intake will be computed and displayed continuously, but achieving an exact fit may still be tedious.

In the case of two component diets, an exact solution to the amounts of forage and compound required can be achieved by using the normal method of solving quadratic equations, as shown by AFRC (1990) in their Appendix 1c.

Prediction of performance

Prediction of animal performance from known intakes of ME also requires that the dry matter intake is known, so that M/D or q_m can be calculated. There are two different approaches, depending on whether the animal is growing or lactating. The equations required have all been quoted earlier, but require some manipulation to achieve the desired result.

Growing animals

Energy retention (R) from a known ME intake, I $(= M_{mp}/E_m)$, for a given value of q_m is given by equation (13) in Chapter One. Since $R = E_f/E_m$, the predicted liveweight gain is then calculated from equations (61) and (37) in Chapter Two, using the appropriate values for the factors C2, C3 and C4. The sequence is:

$$R = B(1 - e^{-kl}) - 1 \tag{13}$$

$$E_f \text{ (MJ/d)} = R \times E_m \qquad \text{rearranged (38)}$$

$$[EV_g] \text{ (MJ/kg)} = \frac{C2(4.1 + 0.0332W - 0.000009W^2)}{(1 - C3 \times 0.1475\Delta W)} \tag{61}$$

$$\Delta W \text{ (kg/d)} = E_f/(C4 \times [EV_g]) \qquad \text{rearranged (37)}$$

Since ΔW is also a term in equation (61), the two equations must be combined, giving:

$$\Delta W \text{ (kg/d)} = E_f/C4(X + 0.1475E_f) \tag{153}$$

where $X = C2(4.1 - 0.0332W - 0.000009W^2)$ taken from equation (61).

Lactating animals

Equation (17) in Chapter One can be used for performance prediction by appropriate manipulation, since all the parameters in it will be known except that which is to be predicted. Replacing E/k by M throughout:

$$M_{mp} \text{ (MJ/d)} = C_L\{M_m + M_l + M_g + M_c\} \tag{17}$$

To predict milk yield at a known or assumed value for liveweight change (ΔW), this equation is solved for M_l, using:

$$M_m \text{ (MJ/d)} = C1(F + A)/k_m \tag{39}$$

$$C_L = 1 + 0.018(L - 1), \text{ where } L = M_{mp}/M_m \tag{12}$$

$$M_g \text{ (MJ/d)} = 19/0.95k_l \text{ if } \Delta W \text{ is +ve, or 16MJ/kg if -ve}$$

$$M_c \text{ (MJ/d)} = E_c/0.133, \text{ where}$$

$$E_c \text{ (MJ/d)} = 0.025W_c(E_t \times 0.0201e^{-0.0000576t}) \text{ and} \tag{71}$$

$$\log_{10}(E_t) = 151.665 - 151.64e^{-0.0000576t} \tag{70}$$

$$[EV_l] \text{ (MJ/kg)} = 0.0384[BF] + 0.0223[P] + 0.0199[La] - 0.108 \tag{53}$$

$$Y \text{ (kg/d)} = (M_l \times k_l)/[EV_l], \text{ which is equation (56) rearranged.}$$

To predict liveweight change, the following sequence is used:

$$[EV_l] \text{ (MJ/kg)} = 0.0384[BF] + 0.0223[P] + 0.0199[La] - 0.108 \tag{53}$$

$$M_l \text{ (MJ/d)} = (Y \times [EV_l])/k_l \qquad (56)$$

Equation (17) is rearranged to become:

$$M_g \text{ (MJ/d)} = (M_{mp}/C_L) - M_m - M_l - M_c \qquad (17)$$

Using equation (39) to calculate M_m, equation (12) for C_L and equations (70) and (71) for M_c, all listed above, M_g can be calculated by difference. Liveweight change (ΔW) is then calculated from M_g, using the appropriate $[EV_g]$ value for ΔW, depending on whether M_g is found to be negative or positive.

Responses to changes in ME intake of lactating animals

The above sequences of calculations assume that either milk yield or liveweight change are known. The system as outlined in this Manual does not deal with responses to additional ME inputs for lactating animals, which requires that the extra ME is partitioned between the synthesis of milk or tissue gain. ARC (1980) does deal briefly with this topic on p93, quoting the work of Blaxter & Boyne (1970). If some assumption is made as to the probable partitioning of additional ME between M_l and M_g, then an estimate of the joint responses can be made. Hulme *et al.* (1986) fitted a curvilinear function to describe the relationship between Net Energy intake above maintenance (E_p) to milk production Net Energy (E_l). Linear segments were then fitted to the curve to enable linear programming techniques to be used in formulating dairy cow diets.

Chapter Five

Dairy Cattle

ME and MP requirements for lactation

There are interactions between feed intake, milk secretion and mobilisation of body reserves in lactating dairy cattle, especially in early lactation. The energy and protein requirement models of AFRC (1990;1992) define the efficiencies of the various pathways, but do not predict how the nutrients are partitioned between maintenance needs, milk synthesis and body tissue requirements. Mathematical models of the changes in milk yield and liveweight, as affected by numerous factors, have been published, and use is made below of the more recent models in calculating tables of requirements for different stages of lactation and levels of milk yield.

The ME requirements of lactating dairy cattle are calculated according to the equations in Chapter Two, and are therefore influenced by the ME concentration of the diet, expressed as either M/D, MJ/kgDM, or as $[ME]/[GE]$, q_m. A mean value for the $[GE]$ of dairy cow diets of 18.8MJ/kgDM has been assumed for the conversion of M/D values to q_m, in the following tables, as discussed in Chapter One. Values for q_m are also quoted, to enable comparisons with tables in other AFRC publications.

Milk composition values are taken from MMB (1992), which gives the mean values for England and Wales (as percent per litre of milk) as 4.06% butterfat, 3.29% protein and 4.55% lactose (39.4, 31.9 and 44.2g/kg milk respectively). Using the Tyrell & Reid (1965) equation recommended by AFRC (1990) (see Chapter Two), this gives an energy value, $[EV_l]$, of 3.0MJ/kg milk.

The calculated MP requirements are independent of dietary ME concentration, although, as shown in Chapter Three, the MP supplied by a diet depends on the plane of nutrition, L, also used in calculating the plane of nutrition correction factor for ME requirements. Example DMI, ME and MP requirements and L values are given in Table 5.1, for a range of liveweight changes as kg/d to cover the different stages of lactation. *A safety margin of 5% has been used in the calculation of both ME and MP requirements.*

Table 5.1: ME (MJ/d) and MP (g/d) requirements of *housed* lactating 600kg dairy cattle at M/D of 11.5MJ/kgDM, or q_m = 0.61.

	Daily milk yield											
	40kg				**35kg**				**30kg**			
$_\Delta$**W**	**DMI**	**ME**	**MP**	**L**	**DMI**	**ME**	**MP**	**L**	**DMI**	**ME**	**MP**	**L**
kg/d	kg	MJ	g	xM	kg	MJ	g	xM	kg	MJ	g	xM
-1.0	*21.4*	*246*	*2020*	*3.8*	19.0	218	1786	3.4	16.6	191	1552	3.0
-0.5	*22.6*	*260*	*2093*	*4.0*	20.2	233	1859	3.6	17.9	205	1625	3.2
0	*23.9*	*275*	*2165*	*4.2*	*21.5*	*247*	*1931*	*3.8*	19.1	220	1697	3.4
+0.5	na	na	na	na	na	na	na	na	20.7	238	1819	3.7
	25kg				**20kg**				**15kg**			
-1.0	14.3	164	1318	2.6	12.0	138	1084	2.2	9.7	111	850	1.8
-0.5	15.5	178	1391	2.8	13.2	152	1156	2.4	10.9	125	922	2.0
0	16.8	193	1463	3.0	14.4	166	1229	2.6	12.1	139	995	2.2
+0.5	18.3	211	1585	3.3	16.0	184	1351	2.9	13.6	157	1117	2.5
	10kg				**5kg**				**Dry, 36 wks preg.***			
-0.5	8.6	99	688	1.6	6.4	74	454	1.2		na		
0	9.8	113	761	1.8	7.6	87	527	1.4	8.8	88	420	1.4
+0.5	11.4	131	883	2.1	9.1	105	649	1.7	11.3	113	543	1.8

Assumptions: Dairy cow of 600kg, producing milk with 39.4g/kg butterfat, 31.9g/kg protein and 44.2g/kg lactose (4.06, 3.29 and 4.55% per litre). No pregnancy allowance for ME or MP included.
* Dry cow requirements are for 36 weeks pregnant, diet M/D of 10MJ/kgDM (q_m = 0.53). See also Table 5.3 for pregnancy requirements for ME and MP. na indicates "not applicable".

The amount of dietary dry matter intake (DMI) required to supply the calculated ME requirement (MER) for a given dietary ME concentration at any particular level of production is easily calculated by converting q_m values to M/D, and since q_m x [GE] = [ME], using:

$$DMI \ (kg/d) = MER/(M/D) \tag{154}$$

The DMI values in Table 5.1 are therefore calculated requirements, not estimates of likely voluntary intake of the total diet, which are in Table 5.2, and with which the DMI values in Table 5.1 should be compared. If the DMI requirements stated exceed the estimates in Table 5.2 (*indicated in italics*), then a higher dietary ME concentration will be needed.

Dry matter appetite prediction

Prediction of the total dry matter intake of dairy cattle has been the subject of numerous research papers. TCORN Report No.5 (AFRC 1990) reviewed many of them and recommended the use of equation VH1 of Vladiveloo & Holmes (1979) for general use, rather than that of ARC (1980). Their recommendation includes diets where forages and concentrates are given together (complete diets or Total Mixed Rations) as well as diets based on hay, straw and well made silage, where the concentrates are fed separately. Equation VH1 is:

$$\text{DMI (kg/d)} = 0.076 + 0.404C + 0.013W - 0.129n + 4.12\log_{10}(n) + 0.14Y \tag{155}$$

where C is kg of concentrate DM
and n is the week of lactation.

The forage intake can be calculated by deducting the concentrate dry matter from the DMI predicted from equation (155). A later report (AFRC 1991b) dealt with the prediction of the voluntary intake of silage by cattle, and their best equation (No.6 in their Table 3) for dairy cattle, weeks 3 to 20 of lactation, was:

$$\text{SDMI (kg/d)} = -3.74 - 0.387C + 1.486(F+P) + 0.0066W_n + 0.0136[\text{DOMD}] \tag{156}$$

where F+P is yield of fat and protein, kg/d, and W_n is the actual weight of the cow in the lactation week, n, under consideration, NOT the mean breed weight, so this parameter is affected by stage of lactation.

The national means for fat and protein concentrations in milk (MMB 1992) are 39.2 and 31.8g/kg milk, so that milk yield, Y, times 0.071 can replace (F+P) in this equation, giving:

$$\text{SDMI (kg/d)} = -3.74 - 0.387C + 0.1055Y + 0.0066W_n + 0.0136[\text{DOMD}] \tag{157}$$

Comparison of predictions for TDMI obtained from equations VH1 and AFRC (1991b) showed that VH1 gave values up to more than 1kg/d higher than AFRC (1991b), for a silage with a typical DOMD of 650g/kgDM. Prediction from week of lactation is more practicable than relying on the actual weight of the cow in the week in question as a measure of stage of lactation effects. The later recommendations of AFRC (1991b) are therefore ignored in this section, and the equation VH1 of Vladiveloo & Holmes (1979) relied upon, although their equation has no factors for silage quality incorporated. Using predicted average milk yield figures for week of lactation based on the equation of Morant & Gnanasakthy (1989) (equation (160) below), for flat rate fed dairy cattle, the values in Table 5.2 are obtained from equation (155).

Chapter Five

Effect of silage quality on DM intake

For diets which are silage based, and where the effects of silage quality are to taken into account, AFRC (1990) suggest that the equations of Lewis (1981) may be used, although they are less accurate overall than equation VH1:

$$I \ (g/kgW^{0.75}) = 0.103[DM] + 0.0516[DOMD] - 0.05[N_a] + 45 \quad (158)$$

$$\text{and} \quad SDMI \ (kg/d) = \{1.068xI - 0.00247(IxC) - 0.00337C2 - 10.9\}W^{0.75}/1000$$

$$+ \ 0.00175Y^2 \quad (159)$$

where   C is concentrate DM intake (g/kgW$^{0.75}$),
[N$_a$] is the ammonia N content (g/kg total N)
and   Y is milk yield (kg/d).

Equations 6-9 of AFRC (1991b) also include a factor for silage [DOMD], with a mean value of 1.2kg silage DM per 100g/kg increase in silage [DOMD]. The effects of silage ammonia content are only mentioned in equation (9) of AFRC (1991b), where an increase of 100g ammonia N, [N$_a$], per kg total N reduces silage DMI by 0.5kg/d.

Table 5.2: Total (TDMI) and silage (SDMI) DM intakes (kg/d) of lactating dairy cows, according to week of lactation and weight (kg/d) of concentrate fed.

Week No.	Milk (kg/d)	Total DMI (kg/d) Concentrate (kg/d) as fed				Silage DMI (kg/d) Concentrate (kg/d) as fed			
		6	8	10	12	6	8	10	12
5	34.6	17.0	17.7	18.4	19.1	11.9	10.8	9.8	8.8
10	33.5	17.5	18.2	18.9	19.6	12.3	11.3	10.3	9.3
15	30.8	17.2	17.9	18.6	19.3	12.0	11.0	10.0	8.9
20	26.9	16.5	17.2	17.9	18.6	11.3	10.3	9.3	8.3
25	22.4	15.6	16.3	17.0	17.7	10.5	9.4	8.4	7.4
30	17.7	14.7	15.4	16.0	16.7	9.5	8.5	7.4	6.4
35	13.4	13.7	14.4	15.1	15.8	8.5	7.5	6.5	5.4
40	9.7	12.8	13.4	14.1	14.8	7.6	6.6	5.5	4.5
45	6.6	11.9	12.6	13.3	14.0	6.7	5.7	4.7	3.7

Values for TDMI are for a 600kg dairy cow and can be adjusted by ± 0.65kg per ± 50kg change in liveweight, by ± 0.40kg/kg concentrate as fed, and ± 0.14kg/kg milk yield, for a specified week of lactation, as equation (155).

Maize silage DM intakes

No equations have been published for the voluntary intake of maize silage by dairy cattle, but Phipps & Wilkinson (1985) reviewed French and American experiments with maize silage which demonstrated that as maize silage DM content increased from 200 to 330g/kg, average maize silage intake by dairy cattle rose from 10.5 to 13.3kgDM/d. For typical UK conditions, maize silage DM content now averages above 300g/kg, so that a *maize silage DM intake of 12kg/d would be a reasonable estimate for lactating dairy cattle.*

Prediction of expected milk yield

Daily milk yield (kg/d) features in most equations for the prediction of the dry matter intake of dairy cows. The prediction of expected daily milk yield at a specified stage of the lactation, usually described as lactation week number (n), or number of days since calving (t), has been the subject of a number of papers, since the original work of Wood (1976), who worked with National Milk Records data for cows mostly fed according to yield. Such prediction equations play a vital role in monitoring dairy herd milk production under a quota regime.

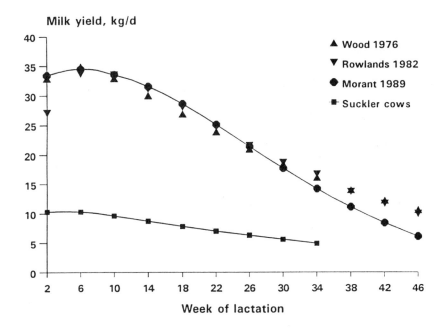

Fig. 5.1: Comparison of lactation curves for dairy and suckler cows.

Morant & Gnanasakthy (1989) reviewed published equations, and derived an improved equation from experimental data for flat rate fed dairy cows. Their equation is:

$$Y \ (kg/d) = \exp\{a - bt1(1 + kt1) + ct1^2 + d/t\} \qquad (160)$$

where t = days since calving (= lactation week, n x 7), and t1 = (t - 150)/100.

Morant & Gnanasakthy's Table 5 gives values for the constants a, b, c, d and k for the prediction of milk yield in the third lactation of cows with lactation yields of about 7000kg. The values given were a = 3.25, b = 0.50, c = 0.0, d = -0.86 and k = 0.39, giving:

$$Y \ (kg/d) = \exp\{3.25 - 0.5t1(1 + 0.39t1) - 0.86/t\} \qquad (161)$$

Adjustments for other levels of lactation milk yield can be made with the parameter "a", which is defined as the natural logarithm ($\log_e$ or ln) of the expected milk yield at day 150 of lactation. Thus for a = 3.25, milk yield at day 150 is 25.8kg/d, whilst 20kg/d gives a value for "a" of 3.0.

Graphical comparison with the equations of Wood (1976), and Rowlands *et al.* (1982) are shown in Fig. 5.1. The equation of Morant & Gnanasakthy (1989) is preferred for predicting the daily milk yield of lactating dairy cows.

Diet formulation for lactating dairy cattle

An example diet for dairy cattle was given in Chapter Four, but some additional examples are given here, making use of dairy compound feeds typical of those being produced in the UK at the time of publication (1993):

Example dairy cow diet 1:

Assumptions: 600kg cow, losing 0.5kg/d and giving 30kg milk with 4.06% fat and 3.29% protein per litre.
MER = 205MJ, L = 3.2, MPR = 1625g, r = 0.08

Feed name	Fresh wt,kg	DM kg	ME MJ	FME MJ	CP g	ERDP g	DUP g	MP g
Grass silage	42.0	10.5	116	85	1491	956	189	
Dairy compound*	8.0	7.0	91	72	1435	861	396	
Totals	50.0	17.5	207	157	2926	1817	585	1686
Requirements		17.9	205	164		1743	514	1625

* *Dairy compound is high ME, 13MJ/kgDM, 18% protein, 7% oil, 7% fibre.*

Diet calculation checks:	ME intake is adequate and within DM intake limit
	M/D = 11.8MJ/kgDM, CP = 167g/kgDM
	y = 11.0g/MJ FME (from L = 3.2)
	MCP = 157 x 11.0 = 1727g/d from FME supply
	MCP = 1817g/d from ERDP supply
	so FME is limiting MCP supply in this diet
therefore	**MPS** = 1727 x 0.6375 + 585 = **1686g/d**
	MPR = **1625**, indicating a surplus of 61g/d

Dairy cow diets based on grass silage commonly have a surplus of ERDP, so that FME supply limits MCP synthesis in the rumen. If inadequate DUP is supplied by the concentrate feed, then total MP supply may be too low, leading to reduced milk protein levels. The addition of good sources of FME such as cereals, fodder beet or molasses to such a diet, and the use of other forages such as maize silage and whole crop cereal, will have beneficial effects upon MCP supply, and therefore MP supply, sufficient to raise milk protein levels, as a number of recent experiments have shown (see example 2 for details).

Example dairy cow diet 2:

Assumptions:	600kg cow, losing 0.5kg/d and giving 30kg milk with 4.06% fat and **3.50%** protein per litre.
	MER = 205MJ, L = 3.2, MPR = 1717g, r = 0.08

Feed name	Fresh wt,kg	DM kg	ME MJ	FME MJ	CP g	ERDP g	DUP g	MP g
Grass silage	32.0	8.0	88	65	1136	720	144	
Fodder beet	15.0	3.0	36	35	189	96	45	
Maize gluten feed	1.7	1.5	19	17	311	195	62	
Wheat, treated	2.5	2.0	26	24	206	120	20	
Rapeseed meal	3.5	3.0	36	32	1200	795	234	
Totals	54.7	17.5	204	**174**	3042	1926	505	**1725**
Requirements		17.9	205	175		1914	497	**1717**

Diet calculation checks:	ME intake is adequate and within DM intake limit
	M/D = 11.7MJ/kgDM CP = 174g/kgDM
	y = 11.0g/MJ FME (from L = 3.2)
	MCP = 1914g/d from FME supply
	MCP = 1926g/d from ERDP supply
	so FME is just limiting MCP supply in this diet
therefore	**MPS** = 1914 x 0.6375 + 505 = **1725g/d**
	MPR = **1717**, indicating a small surplus of 8g/d

Note the substantial increase in FME from 157MJ/d in example 1 to 174MJ/d, due to the inclusion of fodder beet and treated wheat, promoting the synthesis of nearly an extra 200g of MCP, and reducing the need for additional DUP in the concentrates. The diet is optimal for both ERDP and DUP, with only small surpluses of both, so that N excretion by the cows is minimised.

Effects of added dietary fat

A number of dairy cattle experiments have examined the effects of supplying additional energy as dietary fat, both protected and unprotected against rumen bacterial action. Increases in milk yield have been observed, but a common finding was a depression in milk protein level in the milk, although milk protein yield was unchanged. The added fat significantly increased ME intake, by 10-15MJ/d, but did not increase FME intake, so that estimated MP supply was unchanged, but supporting a higher level of milk production. It follows that the use of added fat to dairy cow diets requires the addition of extra DUP to the diet, if adverse changes to milk protein percent are to be avoided. In the example 1 above, the high fat content of the dairy compound is compensated by its high [DUP] content, 57g/kgDM, so that 8kg of it supplies nearly 400g DUP.

Maize silage for dairy cattle

Diets containing maize silage, which has a low protein content (98g/kgDM), and high FME content (9.1MJ/kgDM), are often ERDP limited, see example 3:

Example dairy cow diet 3:

Assumptions: 600kg cow, losing 0.5kg/d and giving 30kg milk with 4.06% fat and 3.29% protein per litre.
MER = 205MJ, L = 3.2, MPR = 1625g, r = 0.08

Feed name	Fresh wt,kg	DM kg	ME MJ	FME MJ	CP g	ERDP g	DUP g	MP g
Maize silage	33.0	10.0	112	91	980	650	140	
Treated straw	3.5	3.0	27	26	105	60	6	
Rapeseed meal	2.3	2.0	24	22	800	530	156	
Soyabean meal	1.7	1.5	20	19	746	393	296	
Maize balancer *	2.3	2.0	25	20	570	334	166	
Totals	42.8	18.5	208	**178**	3201	1967	764	**2012**
Requirements		18.6	205	179		1958	394	**1625**

* Balancer compound contains 12.7MJ/kgDM, 25% protein, 6% oil, 8% fibre.

Diet calculation checks: ME intake is adequate and is within DM appetite
M/D = 11.3MJ/kgDM, CP = 173g/kgDM
y = 11.0g/MJ FME (from L = 3.2)
MCP = 1958g/d from FME supply
MCP = 1967g/d from ERDP supply
so FME is just limiting MCP supply in this diet
therefore **MPS** = 1958 x 0.6375 + 764 = **2012g/d**
MPR = 1625, indicating a surplus of 387g/d

The 387g surplus of MP clearly comes from the DUP supply, which could be reduced if the amounts of some of the protein supplements were reduced and a source of ERDP alone was introduced, such as urea, combined with a source of FME such as barley. Example 4 shows the consequences of so doing. *However, as the response in milk Net Protein to additional MP is 0.2g/gMP, the diet above could raise milk protein by 0.2% (see p66).*

Example dairy cow diet 4:

Assumptions: 600kg cow, losing 0.5kg/d and giving 30kg milk with 4.06% fat and 3.29% protein per litre.
MER = 205MJ, L = 3.2, MPR = 1625g, r = 0.08

Feed name	Fresh wt,kg	DM kg	ME MJ	FME MJ	CP g	ERDP g	DUP g	MP g
Maize silage	40.0	12.0	134	109	1176	780	168	
Treated straw	2.4	2.0	18	18	70	40	4	
Rolled barley	1.4	1.2	15	15	137	101	22	
Rapeseed meal	2.3	2.0	24	22	800	530	156	
Soyabean meal	1.4	1.2	13	13	497	262	197	
Urea, feed grade	0.1	0.1	0	0	288	230	0	
Totals	47.6	18.5	205	176	2968	**1943**	547	**1781**
Requirements		18.6	205	177		1934	416	**1625**

Diet calculation checks: ME intake is adequate and is within DMI limit
M/D = 11.3MJ/kgDM, CP = 160g/kgDM
y = 11.0g/MJ FME (from L = 3.2)
MCP = 1936g/d from FME supply
MCP = 1943g/d from ERDP supply
so FME is just limiting MCP supply in this diet
therefore **MPS** = 1936 x 0.6375 + 547 = **1781g/d**
MPR = 1625, indicating a surplus of 156g/d

The foregoing examples show how diets can be balanced for MP supply by manipulating the feeds included, by having regard to their characteristics for [ME], [ERDP] and [DUP]. They also raise the interesting question as to whether surpluses or deficits of either ERDP or DUP have effects upon the dairy cow's health and performance.

Responses of lactating dairy cattle to protein supply

The question of adequacy of protein in diets for ruminants has been studied and reported extensively, but mostly against the background of the Digestible Crude Protein (DCP), system in widespread use until the early 1980's. Diets formulated by published DCP standards in the UK averaged about 140g CP/kgDM, and the ARC (1980;1984) protein system confirmed that such a level was adequate. Other protein systems subsequently published, reviewed by Alderman (1987), suggested that 160-170g CP/kgDM was required if the rumen microbial need for N was to be met.

Oldham (1984) reviewed the results of many published dairy cattle experiments, and showed that over the range 100-180g/kgDM of [CP] in the diets, significant positive responses in forage intake, total diet digestibility, and hence milk yield were observed as [CP] levels increased. The effects varied in size but increases of 0.2 to 0.4kg DMI per 1% (10g/kg) increase in dietary [CP] were common, ranging 0.1 to 0.8, depending on whether the basal forage was grass silage or maize silage. Increases in milk yield were in line with expectation from the increases in energy intake recorded. Although in some cases it was recorded that protein sources of low solubility were used, there was little characterisation of either forage or supplement degradabilities.

AFRC (1992) reported the results of a large coordinated dairy cow trial, and the results were statistically analysed by Webster (1992), who added in data from Mayne & Gordon (1985). Both sets of trials involved the feeding of grass silage *ad libitum*, plus fixed levels of concentrate, which varied only in their protein content and N degradability. Webster (1992) showed that there was a highly significant linear relationship between increments of MP supply and milk Net Protein (NP_l), output at a particular level of ME intake. He further showed that the data could be pooled, to give a value for NP_l/MP of 0.20, implying that an additional 100gMP would increase milk net protein yield by 20g. This can be compared with the mean efficiency of utilisation of MP for milk synthesis (k_{nl}) of 0.68 adopted in the MP system. The dietary [CP] concentrations in these experiments varied from 140 to 200g/kgDM, and demonstrate values for k_{nl} varying from 0.5 to 0.9. The relevant Figure 10 from AFRC (1992) is shown as Fig. 5.2.

In terms of the MP system described in this Manual, all those points lying to the left of the solid line with the slope of 0.68 are ERDP deficient diets, since they were formulated to ARC (1980;1984) standards, and are showing responses to additional ERDP supply. The data points to the right of the line are demonstrating a response to additional DUP supply, since the concentrate formulations included soyabean meal or fishmeal, both good sources of DUP.

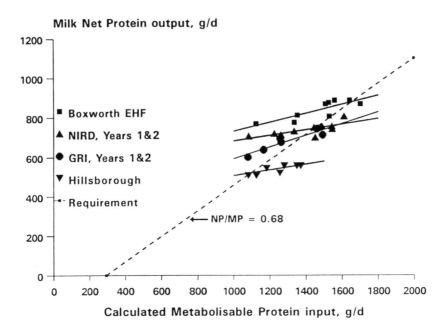

Milk Net Protein output, g/d

■ Boxworth EHF

▲ NIRD, Years 1&2

● GRI, Years 1&2

▼ Hillsborough

- Requirement

←— NP/MP = 0.68

Calculated Metabolisable Protein input, g/d

Fig. 5.2: The relationship between yield of milk protein and MP intake.

Effect of fishmeal on silage intake and digestibility

Fishmeal can have within rumen effects resulting in increases in silage DM intake, diet digestibility and rumen microbial protein synthesis. Most of the published research has been with done with growing beef animals, and is therefore described in Chapter Six, p87. There seems to be no reason why similar effects within rumen should not also be true for dairy cows, and the experiments reported above showed responses in silage DM intake due to fishmeal inclusion which were not explained by the additional DUP supplied. *Higher values for microbial protein synthesis can be used if desired by the user, when fishmeal is added to the diet.*

ME and MP requirements for pregnancy

The ME and MP requirements for pregnancy in cattle are calculated as in Chapters Two and Three, but the amount of weight gain generated by the growing foetus is not normally taken into account, despite the fact that a cow in late lactation may be gaining weight at between 0.5 and 1.0kg/d. ARC (1980), Tables 1.19 and 1.20, give the actual weight of the gravid foetus (W_c), at various stages of gestation, as well as the daily requirements of the foetus for ME (M_c), and MP (MP_c). These are reproduced in Table 5.3, so that the scale of the changes with time can be seen. At week 20, the weight change due to the gravid foetus is only about 0.25kg/d, so that allowance has been made for a further 0.5kg/d of weight gain in the calculation of total ME and MP requirement for the pregnant cow.

ARC (1980) recommends that k_g for lactating animals is $0.95k_l$ (see Chapter One, p3, equation (9)), but that k_f (equation 8) is to be used for non-lactating animals. For a diet M/D of 11MJ/kgDM ($q_m = 0.59$), this implies $k_g = 0.60$ for lactating animals compared to $k_f = 0.46$ for non-lactating. For 0.5kg gain per day, this amounts to 15.8MJ and 20.7MJ of ME respectively.

The ME and MP requirements for the gravid foetus in Table 5.3 are independent of the liveweight of the animal, except in so far as the expected birthweight of the calf is affected by breed type (see Chapter Two, Table 2.3, p30), and adjustments are simply proportional to calf weight, with 40kg being the standard. The data for M_c and MP_c in Table 5.3 therefore define the additional energy and protein requirements of dairy heifers more than 20 weeks in calf. ME required for pregnancy, M_c, is independent of M/D, as $k_c = 0.15$.

Table 5.3: ME (MJ/d) and MP (g/d) requirements of *housed*, pregnant, non-lactating cattle, gaining 0.5kg/d, for M/D of 11MJ/kgDM, or $q_m = 0.59$.

Weeks in calf	Total W_c, kg	ΔW_c, kg/d	M_c, MJ/d	MP_c, g/d	Total requirements* DMI, kg/d	ME, MJ/d	MP, g/d	L
20	11	0.19	2.6	16	7.8	86	432	1.4
24	17	0.27	4.6	28	8.0	88	445	1.4
28	26	0.37	8.0	48	8.3	91	466	1.5
32	38	0.48	13.9	78	8.9	98	497	1.6
36	54	0.60	24.2	121	9.9	109	543	1.8
40	72	0.72	42.0	182	11.7	129	606	2.1

* *Assumptions: Total ME and MP requirements for a dry, pregnant, 600kg cow, gaining 0.5 kg/d liveweight in addition to the gravid foetus, for a 40kg calf.*

Dry matter appetites of pregnant cows

ARC (1980) made no recommendations on the prediction of the dry matter appetites of pregnant cattle, but AFRC (1990) reviewed published work and adopted the equation of Curran *et al.* (1970) for the prediction of forage organic matter intake of dairy cows in late pregnancy, restating it in forage dry matter terms:

$$\text{SDMI (kg/d)} = (0.0003111[\text{DOMD}] - 0.00478C - 0.1102)W^{0.75} \quad (162)$$

where C is concentrate DM intake (kg/d).

Values obtained for total dry matter intake (TDMI), for nil or low levels of concentrate supplementation are given in Table 5.4. Reference to Tables 5.3 and 5.4 shows that for average digestibility silage (650g/kg[DOMD]), dry, pregnant, dairy cattle, normally have a DM appetite in excess of their requirements if fed forage *ad libitum*, unless the diet consists of a single forage, such as grazed grass or silage, which restricts intake either by availability or fermentation characteristics. Dry cows can easily gain too much weight, leading to problems in early lactation, ie the *"fat cow"* or *"fatty liver"* syndrome. Restriction of intake can also be done by raising the cell wall content of the diet, by the inclusion of chopped cereal straw in the diet. A diet M/D of 10MJ/kgDM is quite adequate, as the example dry cow diet shows.

Table 5.4: Total DM intakes (kg/d) of dairy cattle in late pregnancy, according to liveweight (kg), forage [DOMD] (g/kg) and concentrate (kg/d) as fed.

Liveweight (kg)	Concentrate (kg/d as fed)	Forage [DOMD] (g/kg)				
		550	600	650	700	750
500	0	6.4	8.1	9.7	11.4	13.0
	1	6.9	8.5	10.2	11.8	13.4
	2	7.3	8.9	10.6	12.2	13.9
600	0	7.4	9.3	11.2	13.0	14.9
	1	7.7	9.6	11.5	13.4	15.3
	2	8.1	10.0	11.9	13.8	15.6
700	0	8.3	10.4	12.5	14.6	16.8
	1	8.6	10.7	12.8	14.9	17.1
	2	8.9	11.0	13.1	15.2	17.4

Example dry cow diet:

Assumptions: 600kg cow, 36 weeks pregnant and gaining 0.5kg/d. MER = 109, L = 1.8, MPR = 543, r = 0.05

Feed name	Fresh wt,kg	DM kg	ME MJ	FME MJ	CP g	ERDP g	DUP g	MP g
Grass silage	25.0	6.6	73	53	937	594	119	
Barley straw	4.0	3.5	23	22	140	39	56	
Molasses	1.0	0.8	10	10	33	24	0	
Totals	30.0	10.8	106	85	1110	**657**	175	**594**
Requirements		11.2	109	67		833	124	**543**

Diet calculation checks: ME intake is adequate and within DM intake limit
                               M/D = 9.4MJ/kgDM, CP = 103g/kgDM
                               y   = 9.8g/MJ FME (from L = 1.8)
                               MCP = 833g/d from FME supply
                               MCP = 657g/d from ERDP supply
                               so ERDP is limiting MCP supply in this diet
     therefore   **MPS** = 657 x 0.6375 + 175 = **594g/d**
                               MPR = **543g/d**, indicating a surplus of 51g/d

The diet shown meets the energy and protein requirements of a heavily pregnant dairy cow, but the level of [ERDP] in the grass silage does not meet the calculated need to match the [FME] in the diet, and is therefore likely to restrict voluntary intake of the diet as required. Adding or subtracting straw from the diet would easily control feed intake by the joint action of the straw's high cell wall content and further reduction in the ERDP supply to the rumen microbes.

ME and MP requirements of growing dairy heifers

Heifer rearing programmes for heifers to calve down at 2 years of age, weighing about 550kg at parturition, are well established, and require an average liveweight gain of about 0.75kg/d through the rearing period. Attempts to achieve faster rates of gain by using high energy diets (M/D approximately 13MJ/kgDM), have been shown to damage the heifers' ability to milk in the first lactation, so that a restricted range of diet M/D levels will suffice for heifer rearing purposes. The relevant data for Friesian/Holstein heifers are given in Table 5.5.

Inspection of the daily rates of gain due to the gravid foetus given in Table 5.3, column ΔW_c, show that beyond 20 weeks in calf, the allowance for other liveweight gain from Table 5.5 should be reduced to about 0.5kg/d, possibly reducing still further in the last 4 weeks of pregnancy. Over-fatness at calving can lead to metabolic disorders such as *"fatty liver"*, so careful monitoring of feed intakes in late pregnancy is recommended.

Dry matter intakes of growing heifers

The best estimates of the total dry matter intakes of dairy heifers fed grass silage and some concentrates are given by equation (163) and Table 6.3 in Chapter Six.

Table 5.5: ME (MJ/d) and MP (g/d) requirements of Friesian/Holstein heifers.

M/D = 10MJ/kgDM, or q_m = 0.53

			W = 100			W = 200			W = 300		
ΔW kg/d	L xM	y g/MJ	DMI kg	ME MJ	MP g	DMI kg	ME MJ	MP g	DMI kg	ME MJ	MP g
0.50	1.5	9.6	2.8	28	220	4.2	42	261	5.6	56	299
0.75	1.9	9.9	3.5	35	286	5.2	52	322	6.7	67	355

M/D = 11MJ/kgDM, or q_m = 0.59

0.50	1.5	9.5	2.4	26	220	3.7	41	261	4.8	53	299
0.75	1.9	9.9	2.9	32	286	4.4	49	322	5.8	64	355
1.00	2.2	10.2	3.7	40	348	5.4	59	379	7.0	76	409

M/D = 12MJ/kgDM, or q_m = 0.64

0.50	1.5	9.5	2.1	25	220	3.2	39	261	4.3	51	299
0.75	1.8	9.8	2.6	31	286	3.9	46	322	5.0	60	355
1.00	2.1	10.1	3.1	37	348	4.6	55	379	6.0	72	409

M/D = 10MJ/kgDM, or q_m = 0.53

			W = 400			W = 500			W = 600		
ΔW kg/d	L xM	y g/MJ	DMI kg	ME MJ	MP g	DMI kg	ME MJ	MP g	DMI kg	ME MJ	MP g
0.50	1.5	9.5	6.8	68	335	7.9	79	373	8.9	89	411
0.75	1.8	9.8	8.2	82	390	9.5	95	426	10.7	107	465

M/D = 11MJ/kgDM, or q_m = 0.59

0.50	1.5	9.5	5.9	65	335	6.9	75	373	7.8	85	411
0.75	1.8	9.8	7.0	77	390	8.2	90	426	9.2	102	465
1.00	2.1	10.1	8.4	93	441	9.8	108	477	11.0	122	516

M/D = 12MJ/kgDM, or q_m = 0.64

0.50	1.5	9.5	5.2	62	335	6.0	72	373	6.8	82	411
0.75	1.7	9.7	6.1	73	390	7.1	85	426	8.1	97	465
1.00	2.0	10.0	7.2	87	441	8.4	101	477	9.5	114	516

Note: For pregnant heifers beyond 24 weeks of pregnancy, ADD the Mc and MP_c requirements given in Table 5.3.

Chapter Six

Beef Cattle

Growing and finishing cattle

The performance of growing and finishing cattle is largely determined by their voluntary feed intake, so that manipulation of the proportion of concentrate in the total diet is critical in achieving the desired target rate of gain. The setting of the target rate of gain is a complex matter, involving careful budgeting of feed costs over the whole production cycle, and will not be dealt with here, as other authors (Wilkinson 1984; Allen 1990;1992) have dealt with the topic at length.

The energy value of liveweight gains, [EV_g], in cattle vary widely, from below 10MJ/kg for a young weaned bull calf to over 30MJ/kg in finishing heifers over 400kg. One cause of this variation is the proportion of fat in the gains, which for medium castrates varies from 150g/kg at 100kg to 500g/kg at 500kg liveweight. The muscle to bone ratio is also known to vary with both breed and sex of animal, but the protein content of the gain varies relatively little, declining from 170 to 140g/kg for the same weight range. Thus beef cattle generally have low MP requirements compared to dairy cattle or suckler cows, but the ME requirements at high rates of gain are such that the feed costs may exceed the return from the additional gains achieved.

ME and MP requirements of growing and fattening beef cattle

ARC (1980) recommended three breed classes, early, medium and late maturing, when defining the composition of the gains made by growing and finishing cattle, because of the large variation in the energy and protein contents in the liveweight gain at different liveweights. They also recommended additional adjustments ($\pm$ 20-30%) according to the sex of the animal, whether male, castrate or female. Details of these adjustments are in Chapter Two, p28, Tables 2.1 and 2.2 and Chapter Three, Table 3.1. Allen (1992) has given an alternative breed classification (Southgate *et al.*1988) which affects the classification of Friesian cattle, moving them from being a late maturing breed to early maturing. Allen (1992) reduced the breed maturity class and sex adjustments to $\pm$ 20%.

As there are nine combinations of breed type and sex involved, and ME requirements are markedly affected by the ME concentration of the diet (M/D or q_m) offered, requiring large tables, only two typical breed type and sex combinations will be tabulated in this Chapter, Tables 6.1a and b and 6.2a and b. The ME and MP requirements of heifers are given in Chapter Five, p71, Table 5.5. *The usual 5% safety margin has been used in calculating ME and MP requirements of beef cattle that follow.*

Table 6.1a: ME (MJ/d) and MP (g/d) requirements of *housed*, castrate cattle of the medium maturing breeds **(100-300kg W)**.

M/D = 11MJ/kgDM, or q_m = 0.59

			W = 100			W = 200			W = 300		
ΔW kg/d	L xM	y g/MJ	DMI kg	ME MJ	MP g	DMI kg	ME MJ	MP g	DMI kg	ME MJ	MP g
0.50	1.6	9.6	2.4	27	220	3.7	41	261	4.9	54	299
0.75	1.9	9.9	3.0	33	286	4.5	50	322	5.9	65	355
1.00	2.3	10.3	3.8	42	348	5.5	61	379	7.2	79	409

M/D = 12MJ/kgDM, or q_m = 0.64

			W = 100			W = 200			W = 300		
0.50	1.5	9.5	2.1	26	220	3.3	40	261	4.3	52	299
0.75	1.8	9.8	2.6	31	286	4.0	47	322	5.1	62	355
1.00	2.2	10.2	3.2	39	348	4.7	57	379	6.1	74	409
1.25	2.6	10.7	4.0	48	405	5.8	69	432	7.4	89	458

M/D = 13MJ/kgDM, or q_m = 0.69

			W = 100			W = 200			W = 300		
0.50	1.5	9.5	1.9	25	220	2.9	38	261	3.8	50	299
0.75	1.8	9.8	2.3	30	286	3.5	45	322	4.5	59	355
1.00	2.1	10.1	2.8	36	348	4.1	54	379	5.4	70	409
1.25	2.5	10.5	3.4	44	405	4.9	64	432	6.4	83	458
1.50	3.0	10.9	4.2	54	458	5.9	77	481	7.6	99	505

Note: Medium maturing breeds include Lincoln Red, Sussex and Hereford breeds, see Chapter Two, p28, Table 2.2.

Table 6.1b: ME (MJ/d) and MP (g/d) requirements of *housed*, castrate cattle of the medium maturing breeds **(400-600kg W)**.

M/D = 11MJ/kgDM, or q_m = 0.59

			W = 400			W = 500			W = 600		
ΔW	L	y	DMI	ME	MP	DMI	ME	MP	DMI	ME	MP
kg/d	xM	g/MJ	kg	MJ	g	kg	MJ	g	kg	MJ	g
0.50	1.5	9.5	6.0	66	335	7.0	76	373	7.9	87	411
0.75	1.8	9.8	7.2	79	390	8.3	92	426	9.4	104	465
1.00	2.2	10.2	8.7	95	441	10.1	111	477	11.4	125	516

M/D = 12MJ/kgDM, or q_m = 0.64

0.50	1.5	9.5	5.3	63	335	6.1	73	373	6.9	83	411
0.75	1.8	9.8	6.2	75	390	7.3	87	426	8.2	99	465
1.00	2.1	10.1	7.4	89	441	8.6	104	477	9.8	117	516
1.25	2.5	10.5	8.9	107	489	10.4	124	523	11.7	140	563

M/D = 13MJ/kgDM, or q_m = 0.69

0.50	1.5	9.5	4.7	61	335	5.4	71	373	6.2	80	411
0.75	1.7	9.7	5.5	72	390	6.4	83	426	7.3	94	465
1.00	2.0	10.0	6.5	84	441	7.5	98	477	8.5	111	516
1.25	2.4	10.4	7.7	100	489	8.9	116	523	10.1	131	563
1.50	2.9	10.8	9.2	119	533	10.6	138	567	12.0	156	607

Note: Medium maturing breeds include Lincoln Red, Sussex and Hereford breeds, see Chapter Two, p28, Table 2.2.

Table 6.2a: ME (MJ/d) and MP (g/d) requirements of *housed* bulls of the late maturing breeds **(100-300kg W).**

			W = 100			W = 200			W = 300		
ΔW kg/d	L xM	y g/MJ	DMI kg	ME MJ	MP g	DMI kg	ME MJ	MP g	DMI kg	ME MJ	MP g
\multicolumn{12}{c}{M/D = 11MJ/kgDM, or q_m = 0.59}

M/D = 11MJ/kgDM, or q_m = 0.59

ΔW kg/d	L xM	y g/MJ	DMI kg	ME MJ	MP g	DMI kg	ME MJ	MP g	DMI kg	ME MJ	MP g
0.50	1.3	9.2	2.4	26	249	3.7	41	288	4.9	54	324
0.75	1.5	9.5	2.7	30	328	4.2	46	360	5.5	61	392
1.00	1.7	9.7	3.2	35	402	4.8	53	429	6.2	69	456

M/D = 12MJ/kgDM, or q_m = 0.64

ΔW kg/d	L xM	y g/MJ	DMI kg	ME MJ	MP g	DMI kg	ME MJ	MP g	DMI kg	ME MJ	MP g
0.50	1.3	9.2	2.1	25	249	3.3	39	288	4.3	52	324
0.75	1.5	9.5	2.4	29	328	3.7	44	360	4.8	58	392
1.00	1.7	9.7	2.8	33	402	4.2	50	429	5.5	66	456
1.25	2.0	10.0	3.2	38	471	4.8	57	492	6.2	74	515
1.50	2.3	10.3	3.7	44	535	5.4	65	551	7.0	84	571

M/D = 13MJ/kgDM, or q_m = 0.69

ΔW kg/d	L xM	y g/MJ	DMI kg	ME MJ	MP g	DMI kg	ME MJ	MP g	DMI kg	ME MJ	MP g
0.50	1.3	9.3	1.9	24	249	2.9	38	288	3.9	50	324
0.75	1.5	9.5	2.1	28	328	3.3	43	360	4.3	56	392
1.00	1.7	9.7	2.4	32	402	3.7	48	429	4.8	63	456
1.25	1.9	9.9	2.8	36	471	4.2	54	492	5.4	71	515
1.50	2.2	10.2	3.2	41	535	4.7	61	551	6.1	80	571

Note: Late maturing breeds include Charolais, Friesian, Limousin, Simmental and South Devon breeds, as Chapter Two, p28, Table 2.2.

Table 6.2b: ME (MJ/d) and MP (g/d) requirements of *housed* bulls of the late maturing breeds **(400-600kg W)**.

M/D = 11MJ/kgDM, or q_m = 0.59

ΔW kg/d	L xM	y g/MJ	W = 400			W = 500			W = 600		
			DMI kg	ME MJ	MP g	DMI kg	ME MJ	MP g	DMI kg	ME MJ	MP g
0.50	1.3	9.2	5.9	65	359	6.9	76	396	7.9	86	435
0.75	1.5	9.5	6.7	74	425	7.8	86	460	8.8	97	500
1.00	1.7	9.7	7.6	83	486	8.8	97	521	10.0	110	560
1.25	1.9	9.9	8.6	95	543	10.0	110	577	11.4	125	617

M/D = 12MJ/kgDM, or q_m = 0.64

ΔW kg/d	L xM	y g/MJ	DMI kg	ME MJ	MP g	DMI kg	ME MJ	MP g	DMI kg	ME MJ	MP g
0.50	1.3	9.2	5.3	63	359	6.1	74	396	7.0	83	435
0.75	1.5	9.5	5.9	71	425	6.9	82	460	7.8	93	500
1.00	1.7	9.7	6.6	80	486	7.7	93	521	8.7	105	560
1.25	1.9	9.9	7.5	90	543	8.7	105	577	9.9	118	617
1.50	2.1	10.1	8.5	102	596	9.9	119	629	11.2	134	670

M/D = 13MJ/kgDM, or q_m = 0.69

ΔW kg/d	L xM	y g/MJ	DMI kg	ME MJ	MP g	DMI kg	ME MJ	MP g	DMI kg	ME MJ	MP g
0.50	1.3	9.2	4.7	61	359	5.5	71	396	6.2	81	435
0.75	1.5	9.5	5.2	68	425	6.1	79	460	6.9	90	500
1.00	1.6	9.6	5.9	76	486	6.8	89	521	7.7	101	560
1.25	1.8	9.8	6.6	86	543	7.7	100	577	8.7	113	617
1.50	2.1	10.1	7.4	97	596	8.6	112	629	9.8	127	670

Note: Late maturing breeds include Charolais, Friesian, Limousin, Simmental and South Devon breeds, as Chapter Two, p28, Table 2.2.

Dry matter appetite prediction

The voluntary forage dry matter intake functions available for cattle show no influence of breed type or sex. ARC (1980), p65, reviewed published work and concluded that the effects of breed and sex of cattle on feed intake were small, but anticipated that the Meat and Livestock Commission performance trials with beef breeds might give better data, particularly on the Continental breeds. Subsequently, Allen (1992) stated that Continental breeds may consume 10% more feed than British breeds, whilst bulls may consume 10% more than steers.

AFRC (1990) reviewed a number of published prediction equations for the voluntary intake of grass silage by cattle. Their best model (1), requires that the butyric acid content of the silage is known, but model (4) uses ammonia N instead, a commonly available parameter included in commercial silage analyses. Equation (4) has a lower prediction error (0.11 of the mean silage intake) than either the equations of ARC (1980) or Lewis (1981), and is therefore preferred:

SDMI (kg/d) =

$$W^{0.75}(24.96 - 539.7C + 0.108[TDM] - 0.0264[N_a] + 0.0458[DOMD])/1000$$

(163)

> where C is concentrate (gDM/kg $W^{0.75}$),
> N_a is ammonia N (g/kg total N) and
> [TDM] is toluene dry matter (g/kg fresh silage).

The total dry matter intakes of cattle fed average quality grass silage, 250g/kgTDM, 650g/kgDOMD and 100g/kg ammonia in total N are in Table 6.3. Increasing digestibility to 700g/kgDOMD increases silage DM intake by 0.1 to 0.3kg/d as liveweight increases, whilst doubling the ammonia N to 200g/kg decreases silage DM intake by 0.1 to 0.3kg/d when nil concentrate is fed.

Table 6.3: TDMI (kg/d) of beef cattle fed average grass silage and concentrates.

Concentrate as fed (kg/d)	Cattle liveweight (kg)					
	100	200	300	400	500	600
0	2.5	4.2	5.7	7.1	8.4	9.6
1	2.9	4.6	6.1	7.5	8.8	10.0
2	3.3	5.0	6.5	7.9	9.2	10.4
3	na	5.4	6.9	8.3	9.6	10.8
4	na	na	7.3	8.7	9.9	11.2

Note: TDMI values are for grass silage of 250g/kgDM, 650g/kg[DOMD] and 100g/kg ammonia N in total N.
na indicates "not applicable".

No DM intake functions have been published for maize silage, but Allen (1992) gave estimates in his Table 2.1, and indicated a concentrate substitution rate of 0.6kg maize DM for each kg concentrate DM fed. Total Dry Matter Intake (TDMI) figures based on his Table are in Table 6.4.

Table 6.4: TDMI (kg/d) of beef cattle fed maize silage and concentrates.

Concentrate as fed (kg/d)	Cattle liveweight (kg)					
	100	200	300	400	500	600
0	2.7	5.0	6.9	8.4	9.7	na
1	3.1	5.4	7.3	8.8	10.1	na
2	3.5	5.8	7.7	9.2	10.5	na
3	na	6.2	8.1	9.6	10.9	na
4	na	na	8.5	10.0	11.5	na

ARC (1980), Table 2.1, gave total DM intake prediction functions for *coarse* and *fine* diets, both of which have q_m as a factor, as well as metabolic bodyweight and amount of concentrate DM fed, expressed as percent of the diet, C%. *Coarse* diets are defined as those containing long or chopped forages, whilst *fine* diets are those based on concentrates or ground and pelleted forages:

Coarse diets   TDMI $(g/kgW^{0.75})$ = 24.1 + 106.5q_m + 0.37C%   (164)

       where C% is percentage of concentrates in the diet.

Fine diets   TDMI $(g/kgW^{0.75})$ = 116.8 - 46.6q_m (165)

Values for TDMI (kg/d) of cattle fed *coarse* or *fine* diets as defined by ARC (1980) are given in Table 6.5.

Table 6.5: Total Dry Matter Intakes (kg/d) of beef cattle fed *coarse* diets of chopped forage and concentrates, or *fine* diets, according to ARC (1980).

ME concentration		Cattle liveweight, kg					
M/D	q_m	100	200	300	400	500	600
Coarse diets							
7.5	0.4	2.1	3.5	4.8	6.0	7.1	8.1
9.4	0.5	2.4	4.1	5.6	6.9	8.2	9.4
11.3	0.6	2.8	4.7	6.3	7.9	9.3	10.7
13.2	0.7	3.1	5.2	7.1	8.8	10.4	12.0
Per 10% conc.		+0.12	+0.20	+0.27	+0.33	+0.39	+0.45
Fine diets							
9.4	0.5	3.0	5.0	6.7	8.4	9.9	11.3
11.3	0.6	2.8	4.7	6.4	7.9	9.4	10.8
13.2	0.7	2.7	4.5	6.1	7.5	8.9	10.2

Taken from ARC (1980), p61, Table 2.3 - Copyright CAB International.

Diet Formulation for Beef Cattle

An example of diet formulation (but for dairy cattle) was given in Chapter Four, and the same methodical approach is required for beef cattle. However, the use of grass silage as the major part of the diet of growing and fattening cattle, the increasing use of entire bulls for beef production, and the trend to larger, later maturing breeds with higher demands for daily protein deposition, has focused attention on the inadequacies of grass silage "protein" for this purpose.

Unwilted, high digestibility grass silage is characterised by a high [ME] value, > 11MJ/kgCDM, but has a low [FME] value due to its high levels of fermentation acids. Table 4.3 (Chapter Four, p49) would suggest a value of only 7.6MJ [FME]/kgCDM, only 0.69 of [ME]. At an outflow rate of 0.05, the microbial yield (y) will be 10g/MJ of FME. Thus 1kg silage DM will generate about 76gMCP. Because grass silage is highly degradable, it supplies little DUP, values of 20g/kg[DUP] being typical. The MP supply is then 76 x 0.6375 + 20 = 68g per kg silage DM eaten, giving an MP/ME ratio of 6.2. Therefore MP/ME requirement ratios greater than 6.2 indicate that high quality grass silage fed *ad lib*, even if the silage supplies sufficient ME, will supply inadequate MP, resulting in reduced and fattier liveweight gains.

Inspection of the amounts of ME and MP required by growing cattle in Tables 6.4 and 6.5 and the implied MP/ME ratio shows that it varies from 11 for a 100kg bull gaining 1kg/d to 4.9 for a 400kg steer gaining 1.25kg/d. Thus additional MP is required in many situations, particularly with animals below 200kg liveweight, or at high rates of gain for bulls. Additional MP must come from either a supplement rich in DUP, such as fishmeal, or rich in additional FME (barley, swedes), since a lot of silage N is QDP and inefficiently utilised. An example may make this clear:

Example beef cattle diet 1:

Assumptions: 200kg steer, gaining 0.75kg/d.
                      MER = 50MJ, L = 1.8, MPR = 341g, r = 0.05

Feed name	Fresh wt,kg	DM kg	ME MJ	FME MJ	CP g	ERDP g	DUP g	MP g
Grass silage	21.0	4.2	49	37	731	504	88	**319**
Requirements		4.2	47	51		363	110	341

Diet calculation checks: ME intake is adequate and meets DM intake limit

M/D = 49/4.2   = 11.7MJ/kgDM
CP   = 731/4.2   = 174g/kgDM
y   = 9.8g/MJ FME (from L = 1.8)
MCP = 37 x 9.8 = 363g/d from FME supply
MCP = 504g/d from ERDP supply
so FME is limiting MCP supply in this diet

therefore   **MPS** = 363 x 0.6375 + 88
                     = **319g/d**
             MPR = **341g/d,** indicating a small deficit of 22g/d

The diet is both FME and MP limited, due to the limits on silage intake. Although there is a surplus of ERDP, this is due to the low FME supply from the silage, which together with the low [DUP] of silage, result in a shortage of MP for even this modest rate of gain. Higher rates of gain will require the addition of concentrate to supply additional ME, FME and DUP for the higher ME and MP needs. If 1kg of rolled barley is added to the animal's diet, then the calculations are as follows:

Example beef cattle diet 2:

Assumptions: 200kg steer, gaining 1.0kg/d.
                      MER = 54MJ, L = 2.2, MPR = 404g, r = 0.05

Feed name	Fresh wt,kg	DM kg	ME MJ	FME MJ	CP g	ERDP g	DUP g	MP g
Grass silage	18.0	3.6	42	32	626	432	76	
Barley, rolled	1.1	1.0	13	12	114	89	14	
Totals	19.1	4.6	55	**44**	740	521	90	**376**
Requirements		4.6	54	51		449	118	404

Diet calculation checks: ME intake is adequate and meets DM intake limit

M/D = 12.0MJ/kgDM, CP = 161g/kgDM

y = 10.2g/MJ FME (from L = 2.2)

MCP = 449g/d from FME supply

MCP = 521g/d from ERDP supply

so FME is still limiting MCP supply in this diet

therefore   **MPS** = 449 x 0.6375 + 90

= **376g/d**

MPR = **404g/d,** indicating a small deficit of 28g/d

The diet is therefore marginal for MP supply for 1kg gain per day, despite the diet having a [CP] of 161g/kgDM, because barley, despite raising MCP supply by supplying additional FME, is a poor source of DUP, only 14g/kgDM.

One of the best sources of DUP for ruminants is fishmeal, because of the heat treatment to which it has been subjected in processing. The use of fishmeal as the only supplement to high quality grass silage is the basis of the intensive grass silage beef system developed at Rosemaund EHF, as the following example for a 200kg bull will show:

Example beef cattle diet 3 - Rosemaund bull beef grass silage diet:

Assumptions: 200kg bull, gaining 1.25kg/d.

MER = 57MJ, L = 1.9, MPR = 492g, r = 0.05

Feed name	Fresh wt,kg	DM kg	ME MJ	FME MJ	CP g	ERDP g	DUP g	MP g
Grass silage	22.5	4.5	53	40	783	540	95	
Fishmeal	0.45	0.4	6	5	274	120	124	
Totals	22.95	4.9	59	**45**	1057	660	218	**492**
Requirements		5.0	57	67		446	208	502

Diet calculation checks: ME intake is adequate and meets DM intake limit

M/D = 12.0MJ/kgDM, CP = 216g/kgDM

  y   = 9.9g/MJ FME (from L = 1.9)

MCP = 446g/d from FME supply

MCP = 660g/d from ERDP supply

so FME is still limiting MCP supply in this diet

therefore   **MPS** = **446 x 0.6375 + 218**

= **502g/d**

MPR = **492g/d,** indicating a small surplus of 10g/d

The addition of 0.4kg fishmeal DM added over 100g DUP to the diet, without increasing the FME and ME supply appreciably, and in consequence the diet will meet the MP requirements of a fast growing Continental bull, requiring 160g/d more MP than the steer fed grass silage *ad lib* and only gaining 0.75kg/d in example 1. Note, however, that there is a large (200g/d) surplus of ERDP indicated, which will be excreted as urea, and that the [CP] content of the diet is 216g/kgDM. The overall efficiency of protein utilisation is therefore low.

Effect of fishmeal on silage intake and digestibility

The preceding example demonstrated how the amount of DUP supplied by about 0.5kg of fishmeal enabled higher protein gains to be made by a fast growing bull. Significant effects of fishmeal supplementation of grass silage on growing animal performance have been observed in numerous feeding trials, summarised by Miller & Pike (1987). Depending on the class of stock and whether the fishmeal was replacing another concentrate feed or was an addition to the diet, average increases in liveweight gain of 100 to 250g/d were recorded.

Subsequent research (Thomas *et al.* 1980; Gill *et al.* 1987; Dawson *et al.* 1988; Beever *et al.* 1990; Gibb & Baker 1992) has confirmed that fishmeal can have within rumen effects resulting in increases in silage DM intake, diet digestibility and rumen microbial protein synthesis. Beever *et al.*(1990) also reported significant reductions in the fluid outflow from the rumen. These effects are not observed consistently, being affected by frequency of feeding of the fishmeal supplement and whether the silage is restricted fed or *ad lib*. Greater effects have been observed with poorer quality silage and lower rates of gain. It seems probable that fishmeal addition can improve the synchrony of release of energy and N containing nutrients to the rumen microbes, due to the slow rate of degradation of fishmeal, only 0.02/h, comparable to the rate of degradation of the cell wall fraction of the forage.

These within rumen effects of fishmeal addition to grass silage diets lie outside the MP system as presently described, but because of the factorial nature of the system, higher values for microbial protein synthesis can be inserted at the discretion of the user if so desired.

Maize silage for beef cattle

Grass silage as the basis for diets for intensive fattening beef cattle presents special problems, and inefficient utilisation of the feed protein is almost inevitable, except for heifer rearing or store animals, as the previous example diets have shown. The use of maize silage for beef is growing in popularity, but presents a totally different diet formulation problem, as has already been discussed with dairy cow diets.

Example beef cattle diet 4 - maize silage diet:

Assumptions: 200kg bull, gaining 1.25kg/d.
MER = 57MJ, L = 1.9, MPR = 492g, r = 0.05

Feed name	Fresh wt,kg	DM kg	ME MJ	FME MJ	CP g	ERDP g	DUP g	MP g
Maize silage	15.0	4.1	46	37	402	275	53	
Soyabean meal	0.6	0.5	7	6	249	158	73	
Fish meal	0.33	0.3	4	4	206	90	93	
Totals	15.93	4.9	57	**47**	856	522	219	**515**
Requirements		5.0	57	48		465	195	**492**

Diet calculation checks: ME intake is adequate and meets DM intake limit

M/D = 11.6MJ/kgDM, CP = 175g/kgDM
y   = 9.9g/MJ FME (from L = 1.9)
MCP = 465g/d from FME supply
MCP = 522g/d from ERDP supply
so FME is limiting MCP supply in this diet

therefore   **MPS** = 465 x 0.6375 + 219
= **515g/d**
MPR = **492g/d,** indicating a surplus of 23g/d

With heavier cattle, where the MP/ME ratio of their requirements is wider, then appreciably less DUP is required and the inclusion of urea can reduce feed costs appreciably, as the next example shows:

Example beef cattle diet 5 - maize silage diet:

Assumptions: 400kg bull, gaining 1.5kg/d.
MER = 95MJ, L = 1.9, MPR = 596g, r = 0.05

Feed name	Fresh wt,kg	DM kg	ME MJ	FME MJ	CP g	ERDP g	DUP g	MP g
Maize silage	15.0	8.0	90	72	784	536	104	
Rapeseed meal	0.6	0.5	6	5	200	144	29	
Urea, feed grade	0.05	0.05	0	0	144	115	0	
Totals	15.7	8.6	96	77	1128	795	133	619
Requirements		8.6	95	80		762	37	596

Diet calculation checks: ME intake is adequate but exceeds DM intake limit
M/D = 11.2MJ/kgDM, CP = 131g/kgDM
y   = 9.9g/MJ FME (from L = 1.9)
MCP = 762g/d from FME supply
MCP = 795g/d from ERDP supply
so FME is just limiting MCP supply in this diet
therefore   **MPS** = 762 x 0.6375 + 133
= **619g/d**
MPR = **596g/d,** indicating a surplus of 23g/d

Note that this diet is almost optimal for both ERDP and DUP supply, because there are only small surpluses of ERDP and DUP, due to the use of a highly degradable protein supplement, rapeseed meal, combined with a small amount of urea. The consequence is that the dietary [CP] is reduced to 131g/kg. It also follows that the efficiency of feed N utilisation has been maximised and N excretion in faeces and urine minimised.

Cereal beef diets

An intensive beef production system of nearly 40 years standing is *"barley beef"*, or *"cereal beef"* as in MLC publications on beef production. For a detailed description of the system see Allen (1990). Essentially, cattle (usually bulls) are fed *ad lib* a diet consisting of rolled barley, protein supplement, minerals and vitamins, but with access to clean dry barley straw, either as bedding or in racks. Barley has a high [ME] content, 12.8MJ/kgDM, but also a high [FME] content, 12.3MJ/kgDM, so that high levels of MCP production will be supported. Example 6 shows how the diet calculations turn out:

Example beef cattle diet 6 - cereal beef diet:

Assumptions: 400kg bull, gaining 1.5kg/d.
                      MER = 97MJ, L = 2.1, MPR = 596g, r = 0.05

Feed name	Fresh wt,kg	DM kg	ME MJ	FME MJ	CP g	ERDP g	DUP g	MP g
Barley, rolled	7.5	6.4	882	79	730	570	90	
Protein concentrate*	1.3	1.1	13	10	413	276	92	
Totals		7.5	95	89	1142	846	182	721
Requirements		7.5	97	84		899	57	596

* Protein concentrate is 11.5MJ/kgDM, 32% protein, 6% oil and 6% fibre.

Diet calculation checks: ME intake is 2MJ short but meets DM intake limit
                           M/D = 12.6MJ/kgDM, CP = 152g/kgDM
                           y   = 10.1g/MJ FME (from L = 2.1)
                           MCP = 908g/d from FME supply
                           MCP = 846g/d from ERDP supply
                           so ERDP is limiting MCP supply in this diet
        therefore   **MPS** = 846 x 0.6375 + 182
                                = **721g/d**
                           MPR = **596g/d,** indicating a surplus of 125g/d

Suckler cows

The principles of diet formulation for pregnant and lactating dairy cows outlined in Chapter Five also apply to pregnant and lactating suckler cows, but adjusting for levels of peak milk yield of the order of only 8-12kg/d by the fourth week of lactation. Lactations are shorter, about 150 days on average for spring calved cows, although longer for autumn calved cows that continue to suckle at grass. Lactating suckler cows have greater dry matter appetites than growing beef cattle, and can also mobilise body tissues to support lactation. Suckler cows are normally fed to lose liveweight during the winter, once they are in calf, but there are limits to such losses from the animal welfare and milk yield aspects. For these reasons, Allen (1992) recommends not exceeding 0.5kg/d liveweight loss.

The milk yields of suckler cows have been studied by Somerville *et al.*(1983) and Wright & Russel (1987). A lactation curve derived by Somerville *et al.* (1983), using the lactation function of Wood (1976), was as follows:

$$Y \text{ (kg/d)} = 8.0n^{0.121} \times e^{-0.0048n} \tag{166}$$

which is shown graphically in Fig. 5.1 in Chapter Five. Recorded milk composition of suckler cows varies, with a tendency to lower levels of butterfat (34-38g/kg), but with variable protein levels (30-40g/kg), depending on whether the cows are on winter feed or at pasture. A mean value of 32g/kg milk *Crude Protein* has been assumed, resulting in an energy value [EV_l] of 3.0MJ/kg milk, and an MP_l requirement of 45gMP/kg milk, similar to that of dairy cattle.

ME and MP requirements of lactating suckler cows

The ME and MP requirements of typical suckler cows are given in Table 6.6, together with their required DM intakes for a diet M/D of 10MJ/kgDM, and the feeding level (L), which determines outflow rate and microbial protein synthesis. Mean values for feeding level, L = 1.5, outflow rate, r = 0.04, and y = 9.5gMCP/MJ FME are appropriate when formulating diets for suckler cattle.

Table 6.6: ME (MJ/d) and MP (g/d) requirements of *housed* 600kg lactating suckler cows at M/D of 10.0MJ/kgDM, or q_m = 0.53.

			Daily milk yield										
		10kg				8kg				6kg			
ΔW	DMI	ME	MP	L	DMI	ME	MP	L	DMI	ME	MP	L	
kg/d	kg	MJ	g	xM	kg	MJ	g	xM	kg	MJ	g	xM	
-0.5	10.3	103	690	1.6	9.3	93	596	1.4	8.2	82	502	1.3	
0	11.8	118	762	1.8	10.7	107	668	1.6	9.6	96	574	1.5	
+0.5	13.6	136	885	2.1	12.5	125	791	1.9	11.4	114	697	1.8	
		4kg				2kg				Dry, 36 wks preg.*			
-0.5	7.1	71	408	1.1	6.1	61	314	0.9			na		
0	8.6	86	481	1.3	7.5	75	387	1.2	8.8	88	420	1.4	
+0.5	10.3	103	603	1.6	9.3	93	509	1.4	11.3	113	543	1.8	

Assumptions: Suckler cow producing milk with 36g/kg butterfat, 32g/kg **crude** *protein and 50g/kg lactose. Adjust ME requirements by ± 6MJ and MP requirements by ± 19gMP for each 50kg change in mean cow liveweight.*
* No allowance for pregnancy. See Table 5.3 for ME and MP requirements. na indicates "not applicable".

Dry matter intake prediction

AFRC (1990) did not study suckler cow experiments or examine the prediction of dry matter intakes by this class of animal, nor did AFRC (1991b) when it reported on the voluntary intake of cattle. Equation (155) in Chapter Five, that of Vladiveloo & Holmes (1979), has therefore been used, inserting predicted milk yields for the week of lactation derived from the equation of Somerville *et al.* (1983). Because of their greater appetites, suckler cows are normally fed on lower quality silages, or the silages fed with straw in addition, so that dietary M/D values of 10MJ/kgDM are typical, rather than the 11-12MJ/kgDM used for dairy cows. The dry matter intakes from equation (155) have therefore been adjusted downwards in line with the coefficient on DOMD in equation (156) of AFRC (1991b), ie -1.4kg per 100g/kg DOMD reduction. The predicted Total Dry Matter Intakes (TDMI) for nil, 1 and 2kg concentrate per day fed with a silage of only 600g/kg DOMD (9.6MJ ME/kgDM) are in Table 6.7.

Table 6.7: Total (TDMI) and silage (SDMI) DM intakes of lactating suckler cows, according to liveweight, week of lactation and weight of concentrate (kg/d).

Week no.	Milk (kg/d)	Liveweight 500kg Concentrate (kg/d as fed)			Liveweight 600kg Concentrate (kg/d as fed)		
		0	1	2	0	1	2
2	10.3	8.4	8.7	9.1	9.7	10.0	10.4
6	10.3	9.8	10.2	10.5	11.1	11.5	11.8
10	9.6	10.1	10.5	10.8	11.4	11.8	12.1
14	8.7	10.1	10.5	10.8	11.4	11.8	12.1
18	7.8	9.9	10.3	10.6	11.2	11.6	11.9
22	7.0	9.7	10.0	10.3	11.0	11.3	11.6
26	6.3	9.3	9.7	10.0	10.6	11.0	11.3

Note: Silage digestibility assumed to be only 600g/kg[DOMD]. Adjust by +0.14kgDM for each 10g/kg increase in silage [DOMD].

Diet formulation for lactating suckler cows

Because of the lower financial returns compared to dairy cows, diets for suckler cows are necessarily based on the cheapest feeds available, with a heavy emphasis on grass silage, straw and local byproducts. Comparison of the required dry matter intakes indicated in Table 6.6 above, and the predicted voluntary total dry matter intakes in Table 6.7 show that constraints on voluntary feed intake are not likely to be a problem, rather the reverse, when formulating diets for cows suckling their calf. Diets based on below average quality grass silage (600g/kg[DOMD], 9.6MJ/kgDM), with or without small amounts of concentrate feed, are dealt with in the following examples:

Example suckler cow diet 1:

| *Assumptions:* | 600kg cow, giving 10kg milk and losing 0.5kg/d. |
| | MER = 103MJ, L = 1.6, MPR = 690g, r = 0.05 |

Feed name	Fresh wt,kg	DM kg	ME MJ	FME MJ	CP g	ERDP g	DUP g	MP g
Grass silage	32.0	8.0	77	58	1120	784	120	
Barley straw	2.4	2.0	14	13	68	22	32	
Dried beet pulp	1.2	1.0	13	12	103	49	38	
Totals	35.6	11.0	104	**83**	1291	855	190	**698**
Requirements		11.5	103	89		797	182	**690**

Diet calculation checks: ME intake is adequate and within DM intake limit

M/D = 9.4MJ/kgDM, CP = 117g/kgDM
y = 9.6g/MJ FME (from L = 1.6)
MCP = 797g/d from FME supply
MCP = 855g/d from ERDP supply
so FME is limiting MCP supply in this diet

therefore **MPS** = 797 x 0.6375 + 190 = **698g/d**
MPR = **690g/d,** indicating a small surplus of 8g/d

As milk yield declines and liveweight losses cease, the level of concentrate supplementation can be reduced, since by the time milk yield is only 5kg/d grass silage and cereal straw alone will meet the cow's requirements for ME and MP, as example 2 shows:

Example suckler cow diet 2:

| *Assumptions:* | 600kg cow, giving 5kg/d milk and nil liveweight change. |
| | MER = 91MJ, L = 1.4, MPR = 527g, r = 0.04 |

Feed name	Fresh wt,kg	DM kg	ME MJ	FME MJ	CP g	ERDP g	DUP g	MP g
Grass silage	32.0	8.0	77	58	1120	784	120	
Barley straw	2.4	2.0	14	13	68	22	32	
Totals	34.4	10.0	91	**71**	1188	806	152	**573**
Requirements		9.5	91	86		660	102	**527**

Diet calculation checks: ME intake is adequate, but slightly above DMI limit
                             M/D = 9.1MJ/kgDM, CP = 119g/kgDM
                               y = 9.3g/MJ FME (from L = 1.4)
Since FME is limiting **MPS** = 660 x 0.6375 + 152 = **573g/d**
                       MPR = **527g/d,** indicating a small surplus of 46g/d

The situation on turning autumn calving suckler cows out to pasture is strikingly different, even assuming quite modest DM intakes of grass. Spring grass of 75-80D has an [ME] of 12.3MJ/kgDM and a high [FME] of 11.4MJ/kgDM, a 60% increase in [FME] over the 60D silage used as winter feed. The increased intakes of both ME and FME will easily support 1.0kg/d liveweight gain and the production of milk with high (38g/kg) protein content, as recorded by Wright (1992). The calculations are shown in example 3:

Example suckler cow diet 3:

Assumptions: 600kg cow at grass, giving 5kg/d milk and gaining 1kg/d.
                 MER = 123MJ, L = 1.9, MPR = 839g, r = 0.05

Feed name	Fresh wt,kg	DM kg	ME MJ	FME MJ	CP g	ERDP g	DUP g	MP g
Fresh grass	50.0	10.0	123	**114**	1560	1160	220	**940**
Requirements		10.0	120	117		1129	120	**839**

Diet calculation checks: ME intake is adequate and within expected DM intake
                             M/D = 12.3MJ/kgDM, CP = 156g/kgDM
                               y = 9.9g/MJ FME (from L = 1.9)
Since FME is limiting **MPS** = 1129 x 0.6375 + 220 = **940g/d**
                       MPR = **839g/d,** indicating a surplus of 101g/d

It appears that low yielding suckler cows turned out to pasture do not need the highest digestibility grass, and that their pasture stocking rates should be such as to limit daily dry matter intake. However, this would be counter-productive as it would reduce the liveweight gains of the suckler calves.

Diets for dry/pregnant suckler cows

The ME and MP requirements of pregnant suckler cows are the same as those for dairy cattle of the same liveweight, so Table 5.3 for requirements, together with Table 5.4 in Chapter Five for DM intakes should be referred to, and the example dry cow (36 weeks pregnant) diet on p69. Because the efficiencies of utilisation, k_c and k_{nc}, of ME and MP for growth of the foetus are unaffected by diet M/D, then the ME required from diets of lower M/D are only increased slightly, depending on the liveweight gain additional to that of the conceptus.

Chapter Seven

Sheep

The profitability of most sheep enterprises in the UK, whether extensive systems on hill land, or intensive lowland early lamb production systems, is closely related to lamb output. This in turn depends upon the number of lambs produced and their growth rate. The importance of correct ewe nutrition is now widely recognised, but the utilisation of hill and lowland grazing presents problems in assessing feed intake and designing feed supplementation systems. The move to the housing of hill ewes from the later part of pregnancy, lambing and the early lamb suckling phase has also brought a greater degree of control over ewe diets at critical stages of the reproductive cycle. Milk production from dairy sheep or wool production are still a minor part of sheep production systems in the UK at present. As with earlier sections, this Chapter does not set out to advise on the feeding of ewes and lambs, but to set out the energy and protein standards recommended by AFRC TCORN Reports Nos. 5 and 9 (AFRC 1990;1992), together with information on voluntary feed intakes and example diets for each class of sheep. The reader is referred to the MLC publication, *Feeding the Ewe* (MLC 1988) for detailed information on ewe nutrition. *A safety margin of 5% has been used in calculating the ME and MP requirements for sheep.*

Hill and lowland ewes

The ewe's nutrient requirements as presently defined are a function of body weight, body condition and number of lambs carried, and are not influenced by breed of ewe and its environment. With the widespread adoption of housing of pregnant ewes, whether of hill or lowland, the differences in types of feed programmes have narrowed considerably. Whilst hay still features in the feeding of ewes, because of the ease of handling hay bales, the feeding of grass silage to housed ewes, either from clamps or as big bales, has become established practice in many flocks. The correct supplementation of these forages will differ because of their significant differences in [FME], [ERDP] and [DUP] contents.

ME and MP requirements of pregnant ewes

The ME and MP requirements of *housed* pregnant hill and lowland ewes, calculated in accordance with the functions given in Chapters Two and Three, are in Table 7.1 The activity allowances of housed, pregnant ewes were reduced to 0.0054 MJ/kgW by AFRC (1990). If the pregnant ewes are *outdoors*, then this value should be increased to 0.0107MJ/kgW for lowland ewes, and 0.024 MJ/kgW for hill ewes. Total lamb weights (W_o) are calculated as recommended by AFRC (1990), using the equations of Donald & Russell (1970), which give appreciably higher total lamb weights than assumed by ARC (1980) for ewes of over 60kg liveweight. The levels of feeding (L) for a specified week of lactation and number of lambs are fairly constant for each liveweight of ewe, and mean values are quoted at the foot of Table 7.1.

Table 7.1: ME (MJ/d) and MP (g/d) requirements of *housed*, pregnant ewes for M/D of 11 MJ/kgDM, or q_m = 0.59.

			Number of weeks in lamb												
			14			16			18			20			
W kg	W$_o$ kg	n	DMI kg	ME MJ	MP g	DMI kg	ME MJ	MP g	DMI kg	ME MJ	MP g	DMI kg	ME MJ	MP g	
40	3.3	1	0.6	6.7	64	0.7	7.4	68	0.7	8.3	72	0.9	9.5	78	
	5.4	2	0.7	7.4	68	0.8	8.6	74	0.9	10.1	81	1.1	12.0	90	
50	3.9	1	0.7	7.9	72	0.8	8.7	76	0.9	9.8	81	1.0	11.2	88	
	6.4	2	0.8	8.8	77	0.9	10.1	83	1.1	11.9	92	1.3	14.2	103	
60	4.5	1	0.8	9.1	80	0.9	10.0	84	1.0	11.2	90	1.2	12.8	98	
	7.3	2	0.9	10.1	85	1.0	11.6	92	1.2	13.7	102	1.5	16.3	115	
70	5.0	1	0.9	10.2	87	1.0	11.2	92	1.1	12.6	98	1.3	14.4	107	
	8.2	2	1.0	11.4	93	1.2	13.1	101	1.4	15.3	112	1.7	18.3	126	
	9.7	3	1.1	12.0	96	1.3	14.0	106	1.5	16.7	119	1.8	20.3	136	
80	5.5	1	1.0	11.3	94	1.1	12.4	99	1.3	13.9	107	1.4	15.9	116	
	9.0	2	1.1	12.6	100	1.3	14.4	109	1.5	17.0	122	1.8	20.2	137	
	10.8	3	1.2	13.3	104	1.4	15.5	115	1.7	18.5	129	2.0	22.5	148	
Values of L, n = 1			1.3			1.4			1.6			1.8			
		2	1.4			1.6			1.9			2.3			
		3	1.5			1.8			2.1			2.6			

Notes: "n" is the number of lambs. No allowance made for liveweight gain in the ewe except in the fleece. For 50g/d liveweight gain in addition to the gravid foetus, add 2.5MJ of ME and 7g of MP$_g$. Subtract 1MJ of ME and 6g/d of MP$_g$ for a liveweight loss of 50g/d in late pregnancy for ewes in condition score > 3.

Dry matter appetite of pregnant ewes

A major problem with the feeding of pregnant ewes is the changes in voluntary feed intake of ewes in late pregnancy, at a time when nutrient demands are increasing rapidly, often leading to the problem of *"twin lamb disease"*. Estimates of likely total feed DM intake are therefore crucial to the formulation of diets for pregnant ewes. ARC (1980) made no recommendations concerning the dry matter appetite of pregnant ewes, but AFRC (1990) recommended the equation of Neal *et al.* (1985) for the prediction of the intake of hay in mixed diets fed to pregnant ewes. The equation was transformed by AFRC (1990) from an organic matter basis to a dry matter basis to become:

$$\text{HDMI (kg/d)} = C(1.9 - 0.076T - 0.002033[DOMD]) + 0.002444[DOMD] - 0.09565LS + 0.01891W_8 - 1.44 \qquad (167)$$

> *where HDMI* = *intake of hay DM (kg/d),*
> *C* = *concentrate DM (kg/d),*
> *LS* = *litter size,*
> *T* = *week of pregnancy and*
> *W_8* = *liveweight of ewe 8 weeks before lambing (kg).*

For silage, the equations of Lewis (*pers.comm.*) were recommended for use by AFRC (1990). These equations do not incorporate the effects of week of pregnancy, but gives silage DM intakes which are systematically lower than those for hay of equivalent digestibility. The Lewis (*pers.comm.*) equations are:

$$I \text{ (g/kgW)} = 0.0202[DOMD] - 0.0905W - 0.0273N_a + 11.62 \quad (168)$$

and

$$\text{SDMI (kg/d)} = 0.001W\{0.946xI - 0.0204(CxI) + 0.569\} \qquad (169)$$

> *where* N_a = *ammonia nitrogen (g/kg total N)*
> *and* C = *concentrate DM (g/kgW).*

Predicted total DM intakes from the Neal *et al.* (1985) equation for typical hay (600 g/kg [DOMD]), and Lewis for typical silage (650 g/kg[DOMD], and 100g/kg ammonia N in total N), are in Table 7.2. Comparison of the required DMI figures in Table 7.1 with the estimated total DM intakes in Table 7.2, shows that beyond the 17th week of pregnancy the addition of 500g/d of concentrate feed may not be enough to avoid an energy deficit occurring. Most pregnant ewes will lose weight during late pregnancy, so the condition of ewes at this time is therefore critical. MLC (1988) recommend that this is acceptable in ewes in condition score 3 or more at 13 weeks pregnant. A daily liveweight loss of about 50g/day can be allowed without detrimental effects upon either lamb birthweights or lamb viability (Lewis *pers.comm.*).

Table 7.2: Total DM intakes (kg/d) of pregnant ewes fed hay 600g/kg[DOMD] or silage, 650g/kg[DOMD] with nil or 0.5kg/d concentrate as fed.

Ewe W (kg)	Lamb nos.	12		16		20		Silage fed ewes	
		0	0.5	0	0.5	0	0.5	0	0.5
40	1	0.7	1.0	0.7	0.9	0.7	0.8	0.7	1.0
	2	0.6	0.9	0.6	0.8	0.6	0.7		
50	1	0.9	1.2	0.9	1.1	0.9	1.0	0.9	1.1
	2	0.8	1.1	0.8	1.0	0.8	0.9		
60	1	1.1	1.4	1.1	1.3	1.1	1.1	1.0	1.3
	2	1.0	1.3	1.0	1.2	1.0	1.0		
70	1	1.3	1.6	1.3	1.5	1.3	1.3	1.1	1.4
	2	1.2	1.5	1.2	1.4	1.2	1.2		
	3	1.1	1.4	1.1	1.3	1.1	1.1		
80	1	1.4	1.8	1.4	1.6	1.4	1.5	1.2	1.5
	2	1.4	1.7	1.4	1.6	1.4	1.4		
	3	1.3	1.6	1.3	1.5	1.3	1.3		

Hay fed ewes — Number of weeks in lamb: 12, 16, 20.

Diet formulation for pregnant ewes

Two example diet formulations for heavily pregnant ewes are given below:

Example pregnant ewe diet 1:

Assumptions: Hay, fed to housed, 60kg pregnant ewe with twins, 18 weeks in lamb, nil weight change, lamb weight, 7.3kg. MER = 13.7MJ, L = 2.1, MPR = 102, r = 0.05

Feed name	Fresh wt,kg	DM kg	ME MJ	FME MJ	CP g	ERDP g	DUP g	MP g
Meadow hay	0.7	0.5	4.6	4.3	41	22	12	
Ewe compound*	0.7	0.6	7.4	6.1	117	79	24	
Totals	1.4	1.1	12.0	10.4	158	**101**	36	**100**
Requirements		1.1	13.7	10.2		103	33	**102**

* *Pregnant ewe compound is 12.4MJ/kgDM, 17% protein, 6% oil, 8% fibre.*

Diet calculation checks: M/D = 10.9MJ/kgDM, CP = 144g/kgDM
y   = 9.9g/MJ FME (from L = 1.9)
MCP = 103g/d from FME supply
MCP = 101g/d from ERDP supply
MPS = 101 x 0.6375 + 36 = **100g/d**

Although the MP needs are met within the DM appetite of the ewe, there is an energy deficit of 1.7MJ/d, suggesting that some liveweight loss may occur.

Example pregnant ewe diet 2:

Assumptions: Grass silage being fed to 60kg pregnant ewe with twins, 18 weeks in lamb, nil weight change, lamb weight 7.3kg..
MER = 13.7MJ, L = 2.1, MPR = 102g, r = 0.05

Feed name	Fresh wt,kg	DM kg	ME MJ	FME MJ	CP g	ERDP g	DUP g	MP g
Grass silage	3.60	0.90	9.9	7.3	128	89	14	
Barley, rolled	0.42	0.36	4.6	4.4	41	32	5	
Fishmeal	0.04	0.04	0.6	0.5	27	12	12	
Totals	4.06	1.30	15.1	12.2	196	133	31	**109**
Requirements		1.30	13.7	13.2		123	29	**102**

Diet calculation checks: M/D = 11.6MJ/kgDM, CP = 151g/kgDM
y   = 10.1g/MJ FME (from L = 2.1)
MCP = 123g/d from FME supply
MCP = 133g/d from ERDP supply
so FME is limiting MCP supply in this diet
MPS = 123 x 0.6375 + 31 = **109g/d**

The concentrate mixture consists of barley with a 10% inclusion of fishmeal to give a concentrate ME of 12.9MJ/kgDM and a CP content of 15% as fed.

ME and MP requirements for lactating ewes

The ME and MP requirements of housed, lactating ewes of different liveweights and varying milk yields, calculated in accordance with the functions given in Chapters Two and Three, and taking [EV$_l$] as 4.7MJ/kg milk, are in Table 7.3. For the typical liveweights of various breeds of ewe, see MLC (1988), Appendix 2. The dry matter intake figures are those required to supply the amount of ME specified for a diet ME concentration of 11.5MJ/kgDM, and are not the DM appetite of ewes. These DMI values should be compared with the voluntary DM intake figures in Table 7.2. Additional ME for increased activity is required for ewes who are outdoors on lowland or hill grazing, where an additional 5kJ/kgW and 19kJ/kgW are required above that of a housed ewe. The amount of additional ME is given as a footnote for each category of ewe liveweight.

Table 7.3: ME (MJ/d) and MP (g/d) requirements of *housed*, lactating ewes for M/D of 11.5MJ/kgDM, q_m = 0.61.

Housed 40kg ewe				Milk yield (kg/d)							
	1.0				2.0				3.0		
ΔW g/d	DMI kg	ME MJ	MP g	L xM	DMI kg	ME MJ	MP g	L xM	DMI kg	ME MJ	MP g	L xM
0	1.2	13.6	133	2.6	1.9	21.9	209	4.1	-	-	-	-
-50	1.0	11.8	127	2.2	1.7	20.0	203	3.8	-	-	-	-
-100	0.9	10.1	121	1.9	1.6	18.2	196	3.4	-	-	-	-

Lowland ewes outdoors + 0.2 MJ/d, ewes on hills + 0.8 MJ/d

Housed 60kg ewe

ΔW	DMI	ME	MP	L	DMI	ME	MP	L	DMI	ME	MP	L
0	1.3	15.6	146	2.1	2.1	23.7	222	3.3	2.8	32.2	297	4.4
-50	1.2	13.8	140	1.9	1.9	22.0	216	3.0	2.6	30.3	291	4.2
-100	1.0	12.1	134	1.7	1.7	20.2	209	2.8	2.5	28.5	285	3.9

Lowland ewes outdoors + 0.3 MJ/d, ewes on hills + 1.1 MJ/d

Housed 80kg ewe

ΔW	DMI	ME	MP	L	DMI	ME	MP	L	DMI	ME	MP	L
0	1.5	17.5	158	1.9	2.2	25.6	234	2.8	2.9	33.9	309	3.7
-50	1.4	15.8	152	1.7	2.1	23.8	228	2.6	2.8	32.0	303	3.5
-100	1.2	14.0	146	1.5	1.9	22.0	221	2.4	2.6	30.2	297	3.3

Lowland ewes outdoors + 0.4 MJ/d, ewes on hills + 1.5MJ/d

Estimation of ewe milk yields

Suitable estimates of likely ewe milk yields published by MLC (1988), adopted by AFRC (1990), are reproduced in Table 7.4.

Table 7.4: Estimates of ewe milk yields (kg/d) by month of lactation.

Litter size	Type of ewe	Month of lactation		
		1	2	3
One lamb	Hill	1.25	1.05	0.70
	Lowland	2.10	1.70	1.05
Two lambs	Hill	1.90	1.60	1.10
	Lowland	3.00	2.25	1.50

Dry matter appetite of lactating ewes

ARC (1980) recommended mean DM intakes over ewe lactations which were a simple function of metabolic liveweight, 80 g/kgW$^{0.75}$ and 85 g/kgW$^{0.75}$ for ewes suckling single and twin lambs respectively. AFRC (1990) preferred equations which took account of amount of concentrate fed, forage type and digestibility, and recommended the following equation for *hay and concentrate* diets for ewes suckling *twin* lambs:

$$\text{TDMI (kg/d)} = \{I - 0.0691(I \times C) + 2.027C\} \times 0.001W \quad (170)$$

$$\textit{where } I \text{ (g/kgW)} = 0.0481[\text{DOMD}] - 5.25 \quad (171)$$
and C is the intake of concentrate DM (g/kgW).

TDMI of ewes with single or triplet lambs are 0.94 and 1.1 times those from equation (170).

Equations (170) and (171) are apparently based on equations derived by Lewis (*pers.comm.*) for the estimation of the hay intake of pregnant ewes, but multiplied through by 1.33 to adjust for lactating ewes. MLC (1988) gave the total DM intake of lactating ewes fed on hay and concentrates in their 3rd week of lactation, suckling twin lambs as 2.8% of bodyweight:

$$\text{TDMI (kg/d)} = 0.028W \quad (172)$$

Values from both equations are in Table 7.5. Similar functions were recommended by AFRC (1990) for *silage based diets*:

$$\text{TDMI (kg/d)} = 0.001W\{0.946 \times I - 0.0204(I \times C) + 0.65 + C\} \quad (173)$$

$$\textit{where } I \text{ (g/kgW)} = 0.0232[\text{DOMD}] - 0.1041W - 0.0314N_a + 13.36 \quad (174)$$

N_a = *ammonia nitrogen (g/kg total N) and C = intake of concentrate (g/kgW).*

Equation (173) closely resembles equation (169) for pregnant ewes, but with the addition of the term C for concentrate intake, whilst equation (174) appears to be exactly 1.15 times equation (168), that of Lewis (*pers.comm.*) for the silage intake of pregnant ewes. MLC (1988) give TDMI figures for ewes fed *silage and concentrates*:

$$TDMI\ (kg/d) = 0.026W \tag{175}$$

Dry matter intake values for ewes of different liveweight fed average hay (600g/kg [DOMD]) and average grass silage, 250g/kg DM, 650g/kg [DOMD] and 100g/kg ammonia in total N, are in Table 7.5. An increase in hay digestibility to 700g/kg [DOMD] increases hay DM intake at nil concentrate intake by 0.1 to 0.3kg/d as W increases. The effects of silage digestibility are smaller but increases in ammonia N to 200g/kg total N reduce predicted silage intake by 0.1 to 0.2kg/d with nil concentrate.

Table 7.5: Total Dry Matter intakes (kg/d) of lactating ewes fed hay or silage.

Concentrate as fed (kg/d)	Liveweight of ewe (kg)									
	40		50		60		70		80	
	H	S	H	S	H	S	H	S	H	S
0	0.9	1.0	1.2	1.2	1.4	1.4	1.7	1.7	1.9	1.9
0.5	1.1	1.2	1.4	1.4	1.6	1.7	1.8	1.9	2.1	2.1
1.0	1.3	1.4	1.5	1.6	1.8	1.9	2.0	2.1	2.2	2.3
1.5	1.5	1.6	1.7	1.8	1.9	2.1	2.2	2.3	2.4	2.6
MLC (1988)	1.1	1.0	1.4	1.3	1.7	1.6	2.0	1.8	2.2	2.1

Note: H = Hay, 600g/kg [DOMD] and
*   S = Grass silage, 650g/kg [DOMD], 100g/kg ammonia N in total N.*

Diet formulation for lactating ewes

An overview of Tables 7.1, 7.2 and 7.3 above reveals that lactating ewes with twin lambs in early lactation have high ME requirements, with feeding levels (L) comparable to high yielding dairy cows. Like the dairy cow, the ewe has the ability to mobilise body tissue. Depending upon the ewe's bodily condition at lambing, losses of 100-150g/d are supportable. Whilst this mobilisation yields about 2MJ/d for 100g/d loss of liveweight, the yield of MP is only 12g/d. Thus a need for additional MP coming from DUP to meet the high MP needs of lactation might be anticipated. However, high rumen outflow rates of 0.07-0.08/h are predicted, with microbial protein synthesis (y) at 11gMCP/MJ FME. Diets with high [FME] contents may generate sufficient MP from MCP synthesis alone.

Example lactating ewe diet 3:

Assumptions: Forage is hay, being fed to 60kg ewe with twin lambs, giving 2kg milk/d and losing 100g/d liveweight.
MER = 22MJ, L = 3.1, MPR = 210g, r = 0.08

Feed name	Fresh wt,kg	DM kg	ME MJ	FME MJ	CP g	ERDP g	DUP g	MP g
Meadow hay	0.8	0.7	6.4	6.0	57	27	20	
Lactating ewe compound*	1.4	1.2	14.9	12.5	247	158	60	
Totals	2.2	1.9	21.3	18.5	304	**186**	80	**199**
Requirements		1.9	21.7	16.8		204	91	210

* *Ewe compound is 12.4MJ/kgDM, 18% protein, 5% oil, 9% fibre.*

Diet calculation checks: M/D = 11.2MJ/kgDM, CP = 160g/kgDM
y = 11.0g/MJ FME (from L = 3.1)
MCP = 204g/d from FME supply
MCP = 186g/d from ERDP supply
so ERDP is limiting MCP supply in this diet
MPS = 186 x 0.6375 + 80 = **199g/d**
MPR = **210g/d**, indicating a deficiency of 11gMP/d.

Example lactating ewe diet 4:

Assumptions: Forage is average quality grass silage, being fed to 80kg ewe with twin lambs, giving 3kg milk/d and losing 100g/d.
MER = 32MJ, L = 3.5, MPR = 298g, r = 0.08

Feed name	Fresh wt,kg	DM kg	ME MJ	FME MJ	CP g	ERDP g	DUP g	MP g
Grass silage	5.6	1.4	15.4	11.3	199	126	25	
Barley, whole	1.0	0.9	11.5	11.1	103	76	16	
Soyabean meal		0.2	2.7	2.5	99	52	39	
Fishmeal		0.1	1.4	1.2	69	26	34	
Totals		2.6	31.0	**26.1**	470	280	114	**308**
Requirements		2.6	31.9	26.5		292	110	**293**

Diet calculation checks: M/D = 11.9MJ/kgDM, CP = 181g/kgDM
                          y   = 11.2g/MJ FME (from L = 3.5)
                          MCP = 292g/d from FME supply
                          MCP = 280g/d from ERDP supply
                          MPS = 280 x 0.6375 + 114 = **293g/d**
                          MPR = **298g/d**, 5gMP/d short of requirement.

The concentrate mixture in the above example has an [ME] of 13MJ/kgDM, [CP] = 226, [ERDP] = 143, [DUP] = 80g/kgDM, giving a high [DUP]/[CP] ratio of 0.35, due to the inclusion of soyabean and fishmeal.

Growing and fattening lambs

A considerable proportion of the fat lamb production in the UK is fattened at grass, when energy intakes are difficult to assess, although manipulation of stocking rate and grass height on offer greatly influence the DM intake of grazing lambs. However, a significant proportion of the annual lamb crop is housed and finished on conserved forages or intensive cereal based diets, where proper diet formulation can affect profitability considerably.

Breed and sex corrections

ARC (1980), followed by AFRC (1990), did not suggest any adjustment for breed of lamb or early or late maturing characteristics in respect of either ME or MP requirements, but did exclude Merino type sheep breeds from consideration. The effects of sex of lamb were reflected in both the maintenance energy requirements and the energy value of liveweight gains. Intact male lambs have M_m values increased by a factor of 1.15, after calculating their fasting metabolism and activity allowances, as indicated in Chapter Two. Separate linear equations (63-65) are given for the EV_g of gains in growing lambs. For the net protein content of gains, $[NP_g]$, castrates and intact males are combined, giving only two equations (97 and 98). The net effect is that gains in intact male lambs have lower $[EV_g]$ values, due to less fat and more protein in the gains than females. AFRC (1990) concluded that the sex corrections for female lambs were inadequate, but suggested no correction factor.

ME and MP requirements of growing and fattening lambs

The ME and MP requirements of housed, growing and fattening lambs calculated in accordance with the equations listed in Chapters Two and Three are given in Tables 7.6, 7.7 and 7.8. Separate tables for each sex of lamb are required because of the specified effects upon both maintenance and production needs for ME and MP. For lambs at grass, the activity allowance should be increased by 5kJ/kgW per day. *The usual 5% safety margin has been added to calculated allowances.*

Table 7.6: ME (MJ/d) and MP (g/d) requirements of *housed*, **female** lambs for maintenance and liveweight gain.

M/D = 10MJ/kgDM, or q_m = 0.53

ΔW g/d	L xM	y g/MJ	W = 20 DMI kg	ME MJ	MP g	W = 30 DMI kg	ME MJ	MP g	W = 40 DMI kg	ME MJ	MP g	W = 50 DMI kg	ME MJ	MP g
50	1.4	9.3	0.5	4.8	47	0.7	6.6	54	0.8	8.3	60	1.0	9.9	66
100	1.8	9.8	0.6	6.2	61	0.9	8.6	66	1.1	10.9	72	1.3	13.2	78
150	2.4	10.4	0.8	8.0	75	1.1	11.1	79	1.4	14.2	83	1.7	17.3	89
200	3.1	11.0	1.0	10.3	88	1.4	14.5	91	1.9	18.7	95	2.3	23.0	100

M/D = 11MJ/kgDM, or q_m = 0.59

ΔW g/d	L xM	y g/MJ	DMI kg	ME MJ	MP g	DMI kg	ME MJ	MP g	DMI kg	ME MJ	MP g	DMI kg	ME MJ	MP g
50	1.4	9.3	0.4	4.6	47	0.6	6.3	54	0.7	8.0	60	0.9	9.5	66
100	1.8	9.8	0.5	5.9	61	0.7	8.1	66	0.9	10.3	72	1.1	12.4	78
150	2.3	10.3	0.7	7.4	75	0.9	10.3	79	1.2	13.1	83	1.4	15.9	89
200	2.9	10.8	0.8	9.2	88	1.2	12.9	91	1.5	16.6	95	1.8	20.3	100

M/D = 12MJ/kgDM, or q_m = 0.64

ΔW g/d	L xM	y g/MJ	DMI kg	ME MJ	MP g	DMI kg	ME MJ	MP g	DMI kg	ME MJ	MP g	DMI kg	ME MJ	MP g
50	1.4	9.3	0.4	4.5	47	0.5	6.1	54	0.6	7.7	60	0.8	9.2	66
100	1.8	9.8	0.5	5.6	61	0.6	7.7	66	0.8	9.8	72	1.0	11.8	78
150	2.2	10.2	0.6	6.9	75	0.8	9.6	79	1.0	12.2	83	1.2	14.8	89
200	2.7	10.7	0.7	8.4	88	1.0	11.8	91	1.3	15.1	95	1.5	18.4	100
250	3.3	11.1	0.9	10.2	102	1.2	14.4	104	1.5	18.6	107	1.9	22.8	111

M/D = 13MJ/kgDM, or q_m = 0.69

ΔW g/d	L xM	y g/MJ	DMI kg	ME MJ	MP g	DMI kg	ME MJ	MP g	DMI kg	ME MJ	MP g	DMI kg	ME MJ	MP g
50	1.4	9.3	0.3	4.3	47	0.5	5.9	54	0.6	7.4	60	0.7	8.9	66
100	1.7	9.7	0.4	5.4	61	0.6	7.4	66	0.7	9.4	72	0.9	11.3	78
150	2.1	10.1	0.5	6.5	75	0.7	9.1	79	0.9	11.5	83	1.1	13.9	89
200	2.6	10.6	0.6	7.8	88	0.8	10.9	91	1.1	14.0	95	1.3	17.0	100
250	3.1	11.0	0.7	9.3	102	1.0	13.1	104	1.3	16.8	107	1.6	20.5	111
300	3.7	11.4	0.9	11.1	115	1.2	15.6	116	1.5	20.1	118	1.9	24.7	123

Outdoors (MJ)	*+0.1*	*+0.15*	*+0.20*	*+0.25*

Table 7.7: ME (MJ/d) and MP (g/d) requirements of *housed*, **castrate** lambs for maintenance and liveweight gain.

M/D = 10MJ/kgDM, or q_m = 0.53

ΔW	L	y	W = 20			W = 30			W = 40			W = 50		
			DMI	ME	MP	DMI	ME	MP	DMI	ME	MP	DMI	ME	MP
g/d	xM	g/MJ	kg	MJ	g	kg	MJ	g	kg	MJ	g	kg	MJ	g
50	1.4	9.3	0.5	4.8	49	0.6	6.4	56	0.8	8.0	63	0.9	9.5	69
100	1.7	9.7	0.6	6.1	64	0.8	8.2	71	1.0	10.1	77	1.2	12.0	83
150	2.2	10.2	0.8	7.8	80	1.0	10.3	85	1.3	12.7	91	1.5	15.1	97
200	2.7	10.7	1.0	10.0	95	1.3	13.0	100	1.6	15.9	105	1.9	18.9	110

M/D = 11MJ/kgDM, or q_m = 0.59

ΔW	L	y	DMI	ME	MP	DMI	ME	MP	DMI	ME	MP	DMI	ME	MP
50	1.3	9.2	0.4	4.6	49	0.6	6.2	56	0.7	7.7	63	0.8	9.1	69
100	1.7	9.7	0.5	5.8	64	0.7	7.8	71	0.9	9.6	77	1.0	11.4	83
150	2.1	10.1	0.7	7.2	80	0.9	9.6	85	1.1	11.8	91	1.3	14.0	97
200	2.6	10.6	0.8	9.0	95	1.1	11.7	100	1.3	14.5	105	1.6	17.2	110

M/D = 12MJ/kgDM, or q_m = 0.64

ΔW	L	y	DMI	ME	MP	DMI	ME	MP	DMI	ME	MP	DMI	ME	MP
50	1.3	9.2	0.4	4.4	49	0.5	6.0	56	0.6	7.4	63	0.7	8.8	69
100	1.7	9.7	0.5	5.5	64	0.6	7.4	71	0.8	9.2	77	0.9	10.9	83
150	2.0	10.0	0.6	6.8	80	0.8	9.0	85	0.9	11.1	91	1.1	13.2	97
200	2.4	10.4	0.7	8.3	95	0.9	10.8	100	1.1	13.4	105	1.3	15.9	110
250	2.9	10.8	0.8	10.0	110	1.1	13.0	114	1.3	16.0	119	1.6	19.0	124

M/D = 13MJ/kgDM, or q_m = 0.69

ΔW	L	y	DMI	ME	MP	DMI	ME	MP	DMI	ME	MP	DMI	ME	MP
50	1.3	9.2	0.3	4.3	49	0.4	5.8	56	0.6	7.2	63	0.7	8.5	69
100	1.6	9.6	0.4	5.3	64	0.5	7.1	71	0.7	8.8	77	0.8	10.4	83
150	2.0	10.0	0.5	6.4	80	0.7	8.5	85	0.8	10.6	91	1.0	12.5	97
200	2.3	10.3	0.6	7.7	95	0.8	10.1	100	1.0	12.5	105	1.1	14.9	110
250	2.8	10.7	0.7	9.1	110	0.9	11.9	114	1.1	14.7	119	1.3	17.5	124
300	3.2	11.0	0.8	10.8	126	1.1	14.0	129	1.3	17.2	133	1.6	20.4	138

Outdoors (MJ)	*+0.1*	*+0.15*	*+0.2*	*+0.25*

Table 7.8: ME (MJ/d) and MP (g/d) requirements of *housed*, **intact male** lambs for maintenance and liveweight gain.

M/D = 10MJ/kgDM, or q_m = 0.53

			W = 20			W = 30			W = 40			W = 50		
ΔW	L	y	DMI	ME	MP	DMI	ME	MP	DMI	ME	MP	DMI	ME	MP
g/d	xM	g/MJ	kg	MJ	g	kg	MJ	g	kg	MJ	g	kg	MJ	g
50	1.4	9.3	0.5	5.3	49	0.7	7.3	56	0.9	9.2	63	1.1	11.0	69
100	1.7	9.7	0.7	6.7	64	0.9	9.3	71	1.2	11.7	77	1.4	14.1	83
150	2.2	10.2	0.8	8.3	80	1.2	11.6	85	1.5	14.8	91	1.8	17.9	97
200	2.7	10.7	1.0	10.3	95	1.4	14.5	100	1.9	18.6	105	2.3	22.7	110

M/D = 11MJ/kgDM, or q_m = 0.59

50	1.3	9.3	0.5	5.1	49	0.6	7.1	56	0.8	8.9	63	1.0	10.6	69
100	1.7	9.7	0.6	6.4	64	0.8	8.8	71	1.0	11.1	77	1.2	13.4	83
150	2.1	10.1	0.7	7.8	80	1.0	10.8	85	1.2	13.7	91	1.5	16.6	97
200	2.5	10.5	0.9	9.4	95	1.2	13.2	100	1.5	16.9	105	1.9	20.5	110
250	3.1	11.0	1.0	11.4	110	1.5	16.1	114	1.9	20.7	119	2.3	25.4	124

M/D = 12MJ/kgDM, or q_m = 0.64

50	1.3	9.3	0.4	5.0	49	0.6	6.8	56	0.7	8.6	63	0.9	10.2	69
100	1.7	9.7	0.5	6.1	64	0.7	8.4	71	0.9	10.6	77	1.1	12.8	83
150	2.0	10.0	0.6	7.3	80	0.8	10.2	85	1.1	12.9	91	1.3	15.6	97
200	2.4	10.4	0.7	8.8	95	1.0	12.2	100	1.3	15.6	105	1.6	18.9	110
250	2.9	10.8	0.9	10.4	110	1.2	14.5	114	1.6	18.6	119	1.9	22.8	124

M/D = 13MJ/kgDM, or q_m = 0.69

50	1.3	9.3	0.4	4.8	49	0.5	6.6	56	0.6	8.3	63	0.8	9.9	69
100	1.6	9.6	0.4	5.8	64	0.6	8.1	71	0.8	10.2	77	0.9	12.2	83
150	2.0	10.0	0.5	7.0	80	0.7	9.7	85	0.9	12.3	91	1.1	14.8	97
200	2.3	10.3	0.6	8.2	95	0.9	11.4	100	1.1	14.6	105	1.4	17.6	110
250	2.7	10.7	0.7	9.6	110	1.0	13.4	114	1.3	17.1	119	1.6	20.9	124
300	3.1	11.0	0.9	11.1	126	1.2	15.6	129	1.5	20.1	133	1.9	24.5	138

Outdoors (MJ)	*+0.1*		*+0.15*			*+0.2*			*+0.25*					

Dry matter appetite of lambs

ARC (1980) Table 2.1 gave DM intake functions for growing sheep fed *coarse* or *fine* diets, where *coarse* diets are defined as those containing long or chopped forage, whilst *fine* diets are those based on concentrates or ground and pelleted forages. ARC (1980) also gave an equation for the DM intake of lambs fed silage alone. The equations are:

Coarse diets $\quad$ TDMI (kg/d) = $\{104.7q_m + 0.307W - 15.0\}W^{0.75}/1000$ $\quad$ (176)

Fine diets $\qquad$ TDMI (kg/d) = $\{150.3 - 78q_m - 0.408W\}W^{0.75}/1000$ $\qquad$ (177)

Silage alone $\quad$ SDMI (kg/d) = $0.046W^{0.75}$ $\qquad\qquad\qquad\qquad$ (178)

AFRC (1990), p783, tested equations (176) and (177) on data from feeding experiments and found that they gave large errors when predicted and observed TDMI were compared. Equation (178) was accurate, with a mean standard error of only $\pm 1.3g/kgW^{0.75}$, provided formaldehyde treated silages were excluded from the database. In the absence of any better prediction functions, the predicted TDMI and SDMI from these equations are in Table 7.9.

Table 7.9: Total Dry Matter intakes (TDMI) (kg/d) and silage DM intakes (SDMI) (kg/d) of growing lambs fed *coarse*, *fine* or grass silage diets only.

ME concentration		Lamb liveweight (kg)						
M/D	q_m	20	30	40	50	60	70	80
Grass silage only								
na	na	0.44	0.59	0.73	0.86	0.99	1.11	1.23
Coarse diets								
8	0.43	0.34	0.50	0.67	0.84	1.03	1.24	1.45
9	0.48	0.39	0.57	0.75	0.95	1.15	1.37	1.60
10	0.53	0.44	0.64	0.84	1.05	1.27	1.50	1.75
11	0.59	0.50	0.71	0.93	1.16	1.39	1.64	1.89
12	0.64	0.55	0.78	1.02	1.26	1.51	1.77	2.04
13	0.69	0.60	0.85	1.11	1.37	1.63	1.91	2.19
Fine diets								
8	0.43	1.03	1.34	1.60	1.82	2.00	2.14	2.26
9	0.48	0.99	1.29	1.54	1.74	1.91	2.04	2.15
10	0.53	0.95	1.24	1.47	1.66	1.82	1.94	2.04
11	0.59	0.91	1.18	1.41	1.58	1.73	1.84	1.93
12	0.64	0.87	1.13	1.34	1.51	1.64	1.74	1.82
13	0.69	0.83	1.08	1.27	1.43	1.55	1.64	1.70

Note: na indicates "not applicable"

Diet formulation for growing and fattening lambs

Given that a great deal of fat lamb is produced whilst they are at pasture, an examination of the ME and MP balances of grazing lambs is of interest. The following example is based on good quality grazing, [DOMD] = 750-800g/kg, and with a [CP] of 156g/kgDM, typical of early season perennial ryegrass pastures.

Example growing lamb diet 5:

Assumptions: 40kg castrate lamb grazing good pasture, gaining 200g/d.
MER = 13.6MJ, L = 2.4, MPR = 105g, r = 0.05

Feed name	Fresh wt,kg	DM kg	ME MJ	FME MJ	CP g	ERDP g	DUP g	MP g
Fresh pasture	6.0	1.1	13.5	12.5	172	**128**	24	**106**
Requirements		1.1	13.4	12.3		130	23	**105**

Diet calculation checks: M/D = 13.5MJ/kgDM, CP = 172g/kgDM
y = 10.4g/MJ FME (from L = 2.4)
MCP = 130g/d from FME supply
MCP = 128/d from ERDP supply
so ERDP is just limiting MCP supply in this diet
MPS = 128 x 0.6375 + 24 = 106g/d
MPR = 105g/d

The calculations show an excellent balance of FME and ERDP, with just the right amount of DUP to meet the MP needs of the lamb for a gain of 200g/d. A [CP] of about 170g/kgDM is required to achieve this balance, which would not be met with a sward of Italian ryegrass, which has a consistently lower [CP] of around 130g/kgDM at a similar level of digestibility. Nor does aftermath grazing have such high [CP] contents under dry summer weather conditions.

Rearing lambs indoors on silage and barley is common practice, and the next example looks at this type of diet:

Example growing lamb diet 6:

Assumptions: 50kg female lamb gaining 150g/d on silage and barley.
MER = 15.9MJ, L = 2.3, MPR = 89g, r = 0.05

Feed name	Fresh wt,kg	DM kg	ME MJ	FME MJ	CP g	ERDP g	DUP g	MP g
Grass silage, 65D	3.6	0.9	9.9	7.3	128	89	13	
Barley, rolled	0.5	0.4	5.1	4.9	46	36	6	
Totals	4.1	1.3	15.0	12.2	174	**125**	19	**99**
Requirements		1.2	15.9	12.1		126	9	89

Diet calculation checks: M/D = 11.5MJ/kgDM, CP = 134g/kgDM
                        y   = 10.3g/MJ FME (from L = 2.3)
                        ERDP is just limiting MCP synthesis, so that
                        MPS = 125 x 0.6375 + 19 = **99g/d**
                        MPR = **89g/d**, a surplus of 10g/d

The addition of barley makes good use of the surplus ERDP from the grass silage, resulting in high levels of MCP synthesis, so that additional DUP is not needed. Expected DM intake is limiting ME intake and thus the rate of gain achieved. A higher M/D diet is needed to achieve higher rates of gain, which is shown in the next example involving the intensive cereal fattening of lambs.

Example growing lamb diet 7:

Assumptions: 30kg entire male lamb gaining 250g/d on a cereal diet.
                   MER = 13.4MJ, L = 2.7, MPR = 114g, r = 0.08

Feed name	Fresh wt,kg	DM kg	ME MJ	FME MJ	CP g	ERDP g	DUP g	MP g
Barley, rolled	1.00	0.9	11.5	11.1	103	89	17	
Rapeseed meal	0.23	0.2	2.4	2.2	80	53	16	
Totals	1.23	1.1	13.9	13.3	183	**142**	33	**124**
Requirements		1.1	13.4	13.3		142	9	**114**

Diet calculation checks: M/D = 12.7MJ/kgDM, CP = 166g/kgDM
                        y   = 10.7g/MJ FME (from L = 2.7)
                        MCP = 142g/d from FME supply
                        MCP = 142g/d from ERDP supply
                        MPS = 142 x 0.6375 + 33 = **124g/d**
                        MPR = **114g/d**, indicating a surplus of 10g/d

Chapter Eight

Goats

Whilst goats have been kept in the UK for centuries, until about 10 years ago the average herd size was quite small. The establishment of large herds of dairy goats is a recent phenomenon, driven no doubt by the imposition of quotas on milk production from dairy farms within the EEC. As a result, there has been little research in the UK on the nutritional needs of goats, whether reared for milk, meat or fibre. Milk production from goats is well established in France, where goat cheese production has a long history. The French research organisation, INRA, has published feeding standards for goats in its publications on ruminant nutrition (INRA 1978;1988). In the UK, the AFRC's Technical Committee on Responses to Nutrients recognised that its Technical Reviews did not cover goat production, and set up a Goat Nutrition Working Party in 1988 to review published research and information in this field. The Working Party's unpublished Report has been drawn upon (with official permission), in drafting this Chapter. For more practical advice and information on goat farming, readers are referred to Wilkinson & Stark (1987) and Mowlem (1988).

Lactating goats

Dairy goats are capable of producing lactation milk yields of over 1500kg, which for an animal weighing only 60-70kg is a remarkable achievement. Commercial dairy goat herds should be capable of averaging at least 1000kg per lactation. The lactation cycle of the goat is similar to that of the dairy cow (Sutton & Mowlem 1991), rising to a peak at 6-8 weeks, and declining slowly over the next 9-10 months. Liveweight usually falls by about 1kg per week for about the first four weeks, rather as in the case of the dairy cow. Whilst estimates of the amount of energy and protein supplied by such weight loss are inadequate at present, some allowance needs to be made for this when formulating diets for dairy goats. The requirements of goats for ME and MP (as detailed in Chapters Two and Three) are given in the combined tables which follow, *with the addition of the usual agreed safety margin of 5% to both requirements.*

ME and MP requirements for lactation

AFRC (1994) did not use a feeding level correction (C_L) in their calculation of ME requirements of lactating goats, but as AFRC (1990) recommended the use of this correction factor for lactating ewes in addition to dairy cows, it has been used here for the sake of consistency. An activity allowance of 0.0095MJ/kgW has been used for lactating goats, as for housed dairy cattle.

Prediction of the energy value, [EV_l], of goat milk has not been studied, but AFRC (1994) recommend mean values for two main breed types, Anglo-Nubian (A-N) and Sannen/Toggenburg (S/T), as 3.355 and 2.835MJ/kg respectively, based on the use of the Tyrell & Reid (1965) equation on the mean composition of the milks (equation (53) in Chapter Two). However, AFRC (1994) also quote Morant (*pers.comm.*), who has collected data on the average lactation milk yield and composition of the milk of six breeds of goats. Goat milk [EV_l] values and MP_l requirements derived from Morant's data are reproduced in Table 8.1.

Table 8.1: Goat lactation milk yields (kg), milk composition (g/kg), energy values of goats' milk [EV_l] (MJ/kg) and MP_l requirements (g/kg).

		Composition (g/kg)				
Breed of goat	Milk yield (kg)	Fat	Protein	Lactose	[EV_l] (MJ/kg)	MP_l (g/kg)
Anglo-Nubian	681	46.5	35.5	43.4	3.33	47.0
Saanen	904	35.1	28.8	44.8	2.77	38.1
British Saanen	970	37.6	29.2	42.8	2.84	38.6
Toggenburg	672	37.1	28.6	45.8	2.87	37.9
British Toggenburg	1090	37.3	29.6	43.8	2.86	39.2
British Alpine	953	41.1	31.1	43.3	3.03	41.2
Golden Guernsey	820	41.9	31.7	43.3	3.07	42.0

The energy and protein content of weight loss in lactating goats has not been well defined, although energy and nitrogen balances have indicated the amount of the body reserves that can be mobilised over the first 40 days of lactation. AFRC (1994) recommend allowing 4.6MJ of ME/d and 30gMP/d from body reserves for the first month of lactation, based on a nominal liveweight loss of 1kg per week as INRA (1988). The goat's ME and MP requirements in Tables 8.2 and 8.3 are therefore tabulated by month of lactation rather than by the amount of liveweight gain or loss. A diet M/D of 11MJ/kgDM has been used, equivalent to about a 50/50 mixture of grass hay and concentrate.

Table 8.2: ME (MJ/d) and MP (g/d) requirements of *housed*, lactating *Anglo-Nubian* goats for M/D of 11MJ/kg DM, or $q_m = 0.59$.

Milk kg/d	Month 1* DM kg	ME MJ	MP g	L xM	Months 2-3 DM kg	ME MJ	MP g	L xM	Months 4-9 DM kg	ME MJ	MP g	L xM
Goat liveweight 50kg												
1	0.9	10.3	64	1.1	1.4	15.3	95	1.6	1.6	17.2	106	1.8
2	1.5	16.1	114	1.7	1.9	21.2	145	2.2	2.1	23.1	156	2.4
3	2.0	22.1	164	2.3	2.5	27.3	196	2.8	2.7	29.2	206	3.0
4	2.6	28.1	214	2.9	3.0	33.4	246	3.4	-	-	-	-
5	3.1	34.3	264	3.5	3.6	39.7	296	4.0	-	-	-	-
6	3.7	40.6	314	4.1	4.2	46.1	346	4.6	-	-	-	-
Goat liveweight 60kg												
1	1.1	11.7	71	1.1	1.5	16.7	102	1.5	1.7	18.6	113	1.7
2	1.6	17.6	121	1.6	2.1	22.6	152	2.0	2.2	24.5	163	2.2
3	2.1	23.5	171	2.1	2.6	28.6	202	2.5	2.8	30.5	213	2.7
4	2.7	29.5	221	2.6	3.2	34.7	252	3.1	-	-	-	-
5	3.2	35.6	271	3.1	3.7	40.9	302	3.6	-	-	-	-
6	3.8	41.8	321	3.7	4.3	47.2	352	4.1	-	-	-	-
Goat liveweight 70kg												
1	1.2	13.1	77	1.1	1.6	18.1	108	1.5	1.8	19.9	119	1.6
2	1.7	18.9	127	1.5	2.2	24.0	159	1.9	2.3	25.8	169	2.1
3	2.3	24.8	177	2.0	2.7	29.9	209	2.4	2.9	31.8	219	2.5
4	2.8	30.8	227	2.4	3.3	36.0	259	2.8	3.4	37.9	269	3.0
5	3.4	36.9	277	2.9	3.8	42.2	309	3.3	-	-	-	-
6	3.9	43.0	327	3.4	4.4	48.4	359	3.8	-	-	-	-

*Note: * ME requirement reduced by 4.6MJ/d and MP requirement by 30g/d. Milk fat 47g/kg, milk protein 36g/kg, [EV$_l$] = 3.355MJ/kg, MP$_l$ = 47.7g/kg.*

Table 8.3: ME, (MJ/d) and MP (g/d) requirements of *housed*, lactating *Saanen/ Toggenburg* goats for M/D of 11MJ/kg DM, or q_m = 0.59.

Milk kg/d	Month 1*				Months 2-3				Months 4-9			
	DM kg	ME MJ	MP g	L xM	DM kg	ME MJ	MP g	L xM	DM kg	ME MJ	MP g	L xM
Goat liveweight 50kg												
1	0.9	9.4	54	1.0	1.3	14.4	86	1.5	1.5	16.2	96	1.7
2	1.3	14.3	95	1.5	1.8	19.4	126	2.0	1.9	21.3	137	2.2
3	1.8	19.3	135	2.0	2.2	24.4	166	2.5	2.4	26.3	177	2.7
4	2.2	24.4	175	2.5	2.7	29.6	207	3.0	-	-	-	-
5	2.7	29.5	215	3.0	3.2	34.8	247	3.5	-	-	-	-
6	3.2	34.7	256	3.5	3.6	40.1	287	4.0	-	-	-	-
Goat liveweight 60kg												
1	1.0	10.9	61	1.0	1.4	15.8	92	1.4	1.6	17.7	103	1.6
2	1.4	15.7	101	1.4	1.9	20.8	133	1.9	2.1	22.6	143	2.0
3	1.9	20.7	142	1.9	2.3	25.8	173	2.3	2.5	27.7	184	2.5
4	2.3	25.7	182	2.3	2.8	30.9	213	2.7	-	-	-	-
5	2.8	30.8	222	2.7	3.3	36.1	254	3.2	-	-	-	-
6	3.3	36.0	262	3.2	3.8	41.4	294	3.6	-	-	-	-
Goat liveweight 70kg												
1	1.1	12.2	67	1.0	1.6	17.2	99	1.4	1.7	19.0	109	1.5
2	1.6	17.1	108	1.4	2.0	22.1	139	1.8	2.2	24.0	150	1.9
3	2.0	22.1	148	1.8	2.5	27.1	179	2.2	2.6	29.0	190	2.3
4	2.5	27.1	188	2.2	2.9	32.2	220	2.5	3.1	34.1	230	2.7
5	2.9	32.2	229	2.5	3.4	37.4	260	2.9	-	-	-	-
6	3.4	37.3	269	2.9	3.9	42.6	300	3.3	-	-	-	-

*Note: * ME requirement reduced by 4.6MJ/d and MP requirement by 30g. Milk fat 37g/kg, milk protein 29g/kg, [EV$_l$] = 2.835MJ/kg, MP$_l$ = 38.4g/kg.*

Dry matter appetite prediction

INRA (1978) gave a dry matter intake prediction function based on over 10 000 measurements with lactating goats:

$$\text{DMI (g/d)} = 423.2Y + 27.8EBW^{0.75} + 440_\Delta W + 6.75F\% \qquad (179)$$

> *where   EBW = empty-body weight in kg,*
> *F% = proportion of forage as percent,*
> *$_\Delta W$ = liveweight change (kg/d) and*
> *Y = milk yield (kg/d) @ 3.5% fat.*

This equation uses *empty*-body weight, but Alderman (1982), using the functions in the INRA (1978) text, converted the equation to *live*weight and DMI as kg/d rather than g/d, giving:

$$\text{DMI (kg/d)} = 0.42Y + 0.024W^{0.75} + 0.4_\Delta W + 0.7F_p \qquad (180)$$

where F_p is proportion of forage as a decimal.

AFRC (1994), after testing equation (180) on experimental data, adopted Alderman's derived equation, but without the liveweight change correction of $0.4_\Delta W$. INRA (1988), (Table 11.3, p173, of the English text version) gives estimated DM intake figures for the second month of lactation which can be accurately fitted by the following equation:

$$\text{DMI (kg/d)} = 0.062W^{0.75} + 0.305Y \qquad (181)$$

Equation (179) gives a mean value 0.19kg/d below the mean value of the AFRC (1994). Predicted DM intakes of lactating goats at zero liveweight change obtained from equations (180) and (181) are in Table 8.4.

Estimation of goats' milk yield

Williams (1993a, b) studied Saanen dairy goat lactation curves, using 152 sets of records from the NIRD, Shinfield for 1987/88 and 255 sets from four commercial dairy goat farms. He fitted an equation of Morant & Gnanasakthy (1989) to the data, which has the peak yield occurring at week 6, as observed by Badamana *et al.* (1990). Adjusted upwards by one standard deviation on the "a" coefficient to give a lactation yield of 1000kg, Williams (1993) equation is:

$$Y \text{ (kg/d)} = 3.47\exp\{-0.618(1 + t1/2)t1 - 0.0707t1^2 - 1.01t\} \qquad (182)$$

where t = days since parturition and t1 = (t - 150)/100.

This function generates typical milk yields as in Table 8.5.

Table 8.4: DM intake (kg/d) of lactating goats according to milk yield (kg/d).

| Milk (kg/d) | Liveweight (kg) | | | | | |
| | 50 | | 60 | | 70 | |
	INRA'78	INRA'88	INRA'78	INRA'88	INRA'78	INRA'88
0	0.9	1.2	0.9	1.3	1.0	1.5
1	1.3	1.5	1.4	1.6	1.4	1.8
2	1.7	1.8	1.8	2.0	1.8	2.1
3	2.1	2.1	2.2	2.3	2.3	2.4
4	2.6	2.4	2.6	2.6	2.7	2.7
5	3.0	2.7	3.0	2.9	3.1	3.0
6	3.4	3.0	3.5	3.2	3.5	3.3

Note: Forage proportion for INRA (1978) equation taken as 0.6. Lower values would be appropriate for milk yields above 4kg/d.

Table 8.5: Estimated milk yields and DM intakes (kg/d) of 60kg lactating goats.

Lactation week number	2	6	10	14	18	22	26	30	34	38	42
Estimated milk yield	4.8	5.0	4.8	4.4	3.9	3.4	2.8	2.2	1.7	1.2	0.9
Estimated DM intake	2.9	3.0	2.9	2.8	2.6	2.4	2.1	1.9	1.7	1.4	1.3

Note: INRA (1978) equation preferred. Adjust DMI by 0.06kg/d for each 10kg liveweight difference from 60kg.

Diet formulation for lactating goats

According to Mowlem (1988), hay is still widely used as the basal forage for dairy goats, but the feeding of grass and/or maize silage in the larger goat herds is also practised, without significant problems of feed refusal. Straw can also feature in goat diets, provided that straw is available *ad lib* and 25-30% rejection is accepted. Goats are highly selective feeders, rejecting the stemmier portions of the straw, and eating material of higher feed value than the complete forage on offer. Although there has been a tendency to feed milking goats complex, coarse concentrate mixes, these are not essential, and rolled, ground or pelleted concentrates such as cereals, sugarbeet pulp and maize gluten feed, or compound feeds can be fed successfully.

Two example diets will show how the ME and MP requirements in Tables 8.2 and 8.3 can be met within the DM intake limits suggested in Tables 8.4 and 8.5. The higher energy value of milk, [EV_1], of 3.355MJ/kg for Anglo-Nubian goats, compared to only 2.835MJ/kg for Saanen/Toggenburg goats, requires an additional 6MJ/d of ME for a 60kg goat giving 5kg milk/d. A higher dietary M/D for the Anglo-Nubian goats of similar weights is needed, as the DM intake information available does not indicate any breed effects.

Example lactating Sannen/Toggenburg goat diet 1:

Assumptions: Forage is average grass hay, being fed to a *housed*, 60kg S/T goat in 1st month of lactation giving 5kg milk/d.
MER = 30.8MJ, L = 2.7, MPR = 222g, r = 0.08

Feed name	Fresh wt,kg	DM kg	ME MJ	FME MJ	CP g	ERDP g	DUP g	MP g
Grass hay	1.6	1.4	12.9	12.0	113	55	39	
Dairy goat compound.*1.7		1.5	18.8	15.6	309	206	66	
Totals	3.3	2.9	31.7	27.6	422	**261**	105	**271**
Requirements		3.0	30.8	24.3		295	56	222

* Dairy goat compound contains 12.5MJ/kgDM, 17% protein, 5% oil, 9% fibre.

Diet calculation checks: ME intake is adequate and within DM appetite
M/D   = 10.9MJ/kgDM, CP = 146/kgDM
y   = 10.7g/MJ FME (from L = 2.7)
MCP   = 295g/d from FME supply
MCP   = 261g/d from ERDP supply
so ERDP is limiting MCP supply by 35g/d
MPS = 261 x 0.6375 + 105 = **271g/d**
MPR = **222g/d,** giving a surplus of 49gMP/d

Whilst not perfectly balanced, this example diet depends very much on the feeding value of the hay, although the forage proportion is only 0.5 in early lactation for a high milk yield of 5kg/d. An ME contribution of 4.6MJ/d from body reserves has been allowed for, but no MP is needed from body reserves in this example, although 30g/d are allowed for in the tabulated MP requirements.

Using better quality grass and maize silage, and straight feeds instead of compound feed, a suitable diet for 70kg Anglo-Nubian lactating goats would be:

Example lactating Anglo-Nubian goat diet 2:

Assumptions: Forage is grass and maize silage, being fed to *housed*, 70kg
                     A-N goat in 1st month of lactation, giving 5kg milk/d.
                     MER = 36.9MJ, L = 2.9, MPR = 277g, r = 0.08

Feed name	Fresh wt,kg	DM kg	ME MJ	FME MJ	CP g	ERDP g	DUP g	MP g
Grass silage	2.0	0.5	5.5	4.1	71	45	9	
Maize silage	3.0	0.9	10.1	8.1	88	59	13	
Barley, rolled	0.8	0.7	9.0	8.6	80	59	13	
Maize gluten feed	0.6	0.5	6.4	5.8	104	65	21	
Rapeseed meal	0.6	0.5	6.0	5.4	200	133	39	
Totals	7.0	3.1	37.0	**32.0**	543	361	95	**318**
Requirements		3.1	36.9	33.1		346	87	**277**

Diet calculation checks: ME intake meets requirement within DM appetite
                             M/D = 11.9MJ/kgDM, CP = 175g/kgDM
                             y   = 10.8g/MJ FME (from L = 2.9)
                             As FME is limiting MCP synthesis
                             MPS = 346 x 0.6375 + 95 = **315g/d**
                             MPR = **277g/d** giving a surplus of 38g/d

The higher [FME] of this diet, compared to for the hay based diet, means
that the synthesis of an additional 60g/d of MCP is predicted, meeting the higher
needs for milk Net Protein of Anglo-Nubian goats without the inclusion of extra
sources of DUP. *The effects of added fat and extra MP for dairy goats are
expected to be similar to those for dairy cattle (see Chapter Five, p64 and 66).*

Pregnant goats

Because of the lack of published data on the composition of the gravid foetus in
goats, but the length of pregnancy and weight of the new-born kids are close to
those of sheep, the view of AFRC (1994) that the ARC (1980) data for ewes can
be used has been adopted. The relevant equations are (73) and (74) in Chapter
Two and (112) and (113) in Chapter Three. Twin kids of 3.95kg each have been
assumed for dairy goats, as AFRC (1994), so that the requirements for one 4kg
lamb have been increased by a factor of 1.975. For Cashmere x feral goats,
2.75kg each for twin kids has been used, or a factor of 1.375. The increase in
total weight, daily gain, Net and Metabolisable Energy and Net Protein
requirements for the gravid foetus in pregnant dairy goats are given in Table 8.6.

Table 8.6: Increase in weight (kg), energy (MJ/d) and protein (g/d) of the gravid foetus of a pregnant goat with twins, M_c (MJ/d) and MP_c (g/d) requirements.

Week of pregnancy	Total weight (kg)	Gain (g/d)	E_c (MJ/d)	M_c (MJ/d)	NP_c (g/d)	MP_c (g/d)
4	0.30	13	0.02	0.1	1	1
8	0.90	33	0.07	0.5	3	3
12	2.29	70	0.22	1.7	8	9
16	5.04	130	0.56	4.2	19	22
20	9.80	213	1.18	8.8	38	45

Note: Total weight of kids taken as 7.9kg as AFRC (1994). No effects due to breed recognised, and no allowance for additional weight gain of dam included.

ME and MP requirements of pregnant goats

Liveweight gain due to the gravid foetus only becomes significant after the third month of pregnancy (see Table 8.6). AFRC (1994) recommend the addition of 1.7MJ ME/d to allow for liveweight gain in the pregnant goat in addition to that of the gravid foetus. This is sufficient to support about 50g/d additional gain (assuming an $[EV_g]$ of about 20MJ/kg in the adult goat). This gain would also contain about 115g/kg protein, so an additional MP allowance of 10g/d has been included to meet this need for the first three months only. The total ME and MP requirements of a dry, pregnant goat carrying twin kids are in Table 8.7.

Dry matter intakes of adult, non-pregnant and pregnant goats

For adult, non-pregnant goats AFRC (1994) recommend using the *coarse* diet equation of ARC (1980) for predicting the dry matter intake of growing lambs, (equation (176) on p104 of this Manual), multiplied by 1.25, which gives:

$$DMI \ (kg/d) = \{130.9q_m + 0.384W - 18.75\}W^{0.75}/1000 \qquad (183)$$

Estimates obtained from this equation are given in Table 8.8.

AFRC (1994) recommend the adoption of the INRA (1988) estimates of the dry matter intakes of pregnant goats, which are in Table 11.1, p171, of the English text of the latter publication. A 10% reduction in estimated dry matter intake in the last month of pregnancy is recommended. No function was quoted from which the figures were derived, but the following linear equation gives an accurate fit to the published figures for the range of liveweights quoted, 40-80kg. Values obtained are also in Table 8.8.

$$DMI \ (kg/d) = 0.53 + 0.0135W \qquad (184)$$

Chapter Eight

Table 8.7: ME (MJ/d) and MP (g/d) requirements of *housed*, pregnant, non-lactating goats with twin kids for M/D of 11MJ/kgDM, or q_m = 0.59.

Weeks in kid	Liveweight (kg)											
	50				60				70			
	DMI kg	ME MJ	MP g	L xM	DMI kg	ME MJ	MP g	L xM	DMI kg	ME MJ	MP g	L xM
4	1.0	11.5	57	1.2	1.2	12.9	63	1.2	1.3	14.3	70	1.2
8	1.1	11.9	59	1.2	1.2	13.3	66	1.2	1.3	14.7	72	1.2
12	1.2	13.1	66	1.4	1.3	14.6	72	1.3	1.4	15.9	79	1.3
16	1.3	14.0	69	1.5	1.4	15.5	75	1.4	1.5	16.8	82	1.4
20	1.7	19.1	93	2.0	1.9	20.5	100	1.8	2.0	21.9	106	1.8

Note: Dam carrying twin kids (total birthweight 7.9kg), and gaining additional 50g/d liveweight for weeks 4 to 12 ONLY (ie +1.8MJ ME/d and +10gMP/d).

Table 8.8: Dry matter intakes (kg/d) of adult, non-pregnant and pregnant goats.

Liveweight (kg)	Adult, non-pregnant		Pregnant goats	
	M/D = 9	M/D = 11	Months 1-4	Month 5
30	0.71	0.89	0.94	0.84
40	0.94	1.16	1.07	0.96
50	1.19	1.45	1.20	1.08
60	1.44	1.74	1.34	1.21
70	1.71	2.05	1.48	1.33
80	2.00	2.37	1.61	1.45

Diet formulation for pregnant goats

Comparison of the estimates of DM intake in Table 8.8 with the required amounts of DM for a dietary M/D of 11MJ/kgDM in Table 8.7, shows that, as with ewes carrying twins, pregnant goats also require high energy diets in the last month of pregnancy, compared to the earlier months. Two example diets for pregnant goats follow to make this clear.

Example pregnant goat diet 3:

Assumptions: Forage is average quality grass silage plus straw, being fed to *housed*, 70kg goat, 2 months pregnant with twin kids total weight 7.9kg, also gaining 50g/d liveweight.
MER = 14.7MJ, L = 1.2, MP = 72g, r = 0.02

Feed name	Fresh wt,kg	DM kg	ME MJ	FME MJ	CP g	ERDP g	DUP g	MP g
Grass silage	4.0	1.0	11.0	8.1	142	104	11	
Barley straw	0.6	0.5	3.6	3.4	17	5	9	
Totals	4.6	1.5	14.6	**11.5**	159	109	20	**88**
Requirements		1.5	14.7	11.8		106	4	**72**

Diet calculation checks: ME intake is adequate and meets DM appetite
M/D = 9.7MJ/kgDM, CP = 106g/kgDM
y = 9.2g/MJ FME (from L = 1.2)
As FME is just limiting MCP supply
MPS = 106 x 0.6375 + 20 = **88g/d**
MPR = **72g/d,** a surplus of 4gMP/d

The next example is for a goat in the fifth month of pregnancy:

Example pregnant goat diet 4:

Assumptions: Forage is average quality grass silage, being fed to *housed*, 70kg goat, 18 weeks pregnant with twin kids, 7.9kg. Mobilisation of body reserves (-3MJ ME/d) allowed.
MER = 16.4MJ, L = 1.4, MP = 92g, r = 0.02

Feed name	Fresh wt,kg	DM kg	ME MJ	FME MJ	CP g	ERDP g	DUP g	MP g
Grass silage	1.6	0.4	4.4	3.2	57	40	6	
Barley, rolled	0.9	0.8	10.2	9.8	91	71	11	
Rapeseed meal	0.1	0.1	1.2	1.1	40	29	6	
Totals	2.6	1.3	15.8	**14.2**	188	140	23	**107**
Requirements		1.3	16.4	15.2		132	5	**92**

Diet calculation checks: ME intake is adequate and meets DM appetite
M/D = 12.2MJ/kgDM, CP = 145g/kgDM
y = 9.3g/MJ FME (from L = 1.4)
as FME is limiting MCP supply
MPS = 132 x 0.6375 + 23 = 107g/d
MPR = 92g/d, giving a surplus of 15gMP/d

In example 3 above, replacing the silage and straw mix with 1.5kg average hay DM would nearly meet the ME and MP requirements, but the diet would be 50g/d deficient in ERDP, which would be likely to reduce hay consumption below its maximum. In example 4, the mobilisation of about 3MJ/d of ME from maternal body reserves has had to be assumed, since the maximum appetite for a high energy diet has been reached. No protein contribution from maternal reserves appears to be necessary, as the example shows. Clearly the main problem is achieving a high enough intake of ME within the animal's DM appetite, requiring a diet with a 30:70 forage/concentrate ratio. If the ERDP needed by the FME from the barley is then supplied, sufficient MCP is synthesised by the rumen microbes to meet the MP needs of the goat, as defined.

Growing goat kids

AFRC (1994) only found data sets for the changes in the composition of castrate male goat kids over time, so no effect of sex of kid is stated, although data indicating lower fat contents in the bodies of adult females were found. The data for energy and protein content of castrate male kids are described by equations (66) and (67) in Chapter Two and equations (103) and (104) in Chapter Three of this Manual. The ME and MP requirements for growing castrate male kids calculated in accordance with the equations in Chapters Two and Three, using the activity allowance of 0.0067MJ/kgW given for housed, fattening lambs, together with the addition of the usual 5% safety margin, are in Table 8.9. *For Cashmere and Angora goats kept for fibre production, add 0.2MJ/d and 14gMP/d, and 0.5MJ/d and 40gMP/d respectively to the values in Table 8.9.*

Dry matter intake of growing goat kids

AFRC (1994) recommend that the ARC (1980) DM intake functions for growing lambs (equations (176) and (177) in the preceding Chapter) should be used for growing goat kids. Estimates of DM intakes of *coarse* and *fine* diets for lambs were given in Table 7.9, reproduced again here for convenience as Table 8.10, but omitting the values for grass silage obtained with lambs (AFRC 1990).

Table 8.9: ME (MJ/d) and MP (g/d) requirements of *housed*, **castrate male** kids.

ΔW g/d	W = 20kg DMI kg	ME MJ	MP g	L xM	W = 30kg DMI kg	ME MJ	MP g	L xM	W = 40kg DMI kg	ME MJ	MP g	L xM
				M/D = 11MJ/kgDM, or q_m = 0.59								
0	0.4	4.6	23	1.1	0.6	6.3	31	1.1	0.7	7.8	38	1.1
100	0.6	7.0	48	1.6	0.9	9.3	55	1.6	1.1	11.5	61	1.5
200	0.9	10.0	74	2.3	1.2	13.1	79	2.2	1.5	16.1	84	2.2
				M/D = 12MJ/kgDM, or q_m = 0.64								
0	0.4	4.5	23	1.1	0.5	6.1	31	1.1	0.7	7.6	38	1.1
100	0.6	6.7	48	1.6	0.8	8.9	55	1.5	0.9	11.0	61	1.5
200	0.8	9.4	74	2.2	1.0	12.4	79	2.1	1.3	15.3	84	2.1
				M/D = 13MJ/kgDM, or q_m = 0.69								
0	0.3	4.4	23	1.1	0.5	6.0	31	1.1	0.6	7.4	38	1.1
100	0.5	6.4	48	1.5	0.7	8.6	55	1.5	0.8	10.6	61	1.5
200	0.7	8.8	74	2.1	0.9	11.6	79	2.0	1.1	14.3	84	2.0

ΔW g/d	W = 50kg DMI kg	ME MJ	MP g	L xM	W = 60kg DMI kg	ME MJ	MP g	L xM	W = 70kg DMI kg	ME MJ	MP g	L xM
				M/D = 11MJ/kgDM, or q_m = 0.59								
0	0.9	9.3	45	1.1	1.0	10.7	52	1.1	1.1	12.0	58	1.1
100	1.3	13.6	67	1.5	1.5	15.7	73	1.5	1.6	17.7	78	1.5
200	1.8	19.1	89	2.2	2.0	22.0	93	2.2	2.3	24.8	97	2.2
				M/D = 12MJ/kgDM, or q_m = 0.64								
0	0.8	9.0	45	1.0	0.9	10.4	52	1.1	1.0	11.7	58	1.1
100	1.1	13.1	67	1.5	1.3	15.1	73	1.5	1.4	17.0	78	1.5
200	1.5	17.9	89	2.1	1.7	20.6	93	2.1	2.0	23.3	97	2.1
				M/D = 13MJ/kgDM, or q_m = 0.69								
0	0.7	8.8	45	1.1	0.8	10.1	52	1.1	0.9	11.5	58	1.1
100	1.0	12.6	67	1.5	1.1	14.5	73	1.5	1.3	16.3	78	1.5
200	1.3	16.9	89	2.0	1.5	19.5	93	2.0	1.7	22.0	97	2.0

Table 8.10: DM intakes (kg/d) of growing goat kids fed *coarse* or *fine* diets.

ME concentration		Goat kid liveweight (kg)						
M/D	q_m	20	30	40	50	60	70	80
Coarse diets								
9	0.48	0.39	0.57	0.75	0.95	1.15	1.37	1.60
10	0.53	0.44	0.64	0.84	1.05	1.27	1.50	1.75
11	0.59	0.50	0.71	0.93	1.16	1.39	1.64	1.89
12	0.64	0.55	0.78	1.02	1.26	1.51	1.77	2.04
13	0.69	0.60	0.85	1.11	1.37	1.63	1.91	2.19
Fine diets								
9	0.48	0.99	1.29	1.54	1.74	1.91	2.04	2.15
10	0.53	0.95	1.24	1.47	1.66	1.82	1.94	2.04
11	0.59	0.91	1.18	1.41	1.58	1.73	1.84	1.93
12	0.64	0.87	1.13	1.34	1.51	1.64	1.74	1.82
13	0.69	0.83	1.08	1.27	1.43	1.55	1.64	1.70

Diet formulation for goat kids

Goat kids are reared for two types of production systems, either as herd replacements, or for meat. The feeding of the unweaned goat kid lies outside the scope of this Manual, since they are not functioning ruminants, see Mowlem (1988) and AFRC (1994). Herd replacements need to be fed economically, a daily weight gain of 50-100g/d being adequate. Given the estimates of dry matter intake in Table 8.10 for *coarse* diets, forage based diets for goat kids cannot be used if high weight gains for meat production from kids are required.

Example growing goat kid diet 5:

Assumptions: Forage is maize silage fed to a *housed*, 30kg female goat, gaining 100g/d.
MER = 9.3MJ, L = 1.6, MP = 55g, r = 0.05

Feed name	Fresh wt,kg	DM kg	ME MJ	FME MJ	CP g	ERDP g	DUP g	MP g
Maize silage	1.65	0.55	6.2	5.0	47	30	3	
Peas, rolled/ground	0.30	0.25	3.5	3.3	65	48	7	
Totals	1.95	0.80	9.7	8.3	112	**78**	10	**60**
Requirements		0.80	9.3	8.1		80	5	55

Diet calculation checks: ME intake is adequate and is within DM appetite
M/D = 12.1MJ/kgDM, CP = 140g/kgDM
y   = 9.6g/MJ FME (from L = 1.6)
MCP = 80g/d from FME supply
MCP = 78g/d from ERDP supply
so ERDP is just limiting MCP supply
MPS = 78 x 0.6375 + 10 = **60g/d**
MPR = **55g/d**, a surplus of 5gMP/d

Note that the M/D of the diet in this example needs to be 12MJ/kgDM to achieve the modest growth rate of 100g/d, but that only 140g/kg of [CP] is required. High rates of gain (200g/d) by goat kids will require intensive cereal type diets, because higher DM and ME intakes are required.

Example growing goat kid diet 6:

Assumptions: Diet is cereal based, with straw on offer, being fed to *housed*, castrate male 30kg goat kid, gaining 200g/d.
MER = 13.1MJ, L = 2.2, MP = 79g, r = 0.05

Feed name	Fresh wt,kg	DM kg	ME MJ	FME MJ	CP g	ERDP g	DUP g	MP g
Barley straw	0.1	0.10	0.7	0.7	3	1	1	
Barley, rolled	0.9	0.80	10.2	9.8	91	71	11	
Rapeseed meal	0.3	0.25	3.0	2.7	100	72	14	
Totals	1.3	1.15	13.9	**13.2**	194	144	26	**112**
Requirements		1.13	13.1	14.1		135	18	**79**

Diet calculation checks: ME intake exceeds requirement within DM appetite
M/D = 12.1MJ/kgDM, CP = 169g/kgDM
y   = 10.2g/MJ FME (from L = 2.2)
MCP = 135g/d from FME supply
MCP = 144g/d from ERDP supply
so FME is limiting MCP supply
MPS = 135 x 0.6375 + 26 = **112g/d**
MPR = **79g/d,** giving a surplus of 33gMP/d

The higher [FME] of this diet, 11.5MJ/kgDM, compared to only 10.3MJ/kgDM in the previous example, calls for higher levels of [CP], 169g/kg, coming from a rapidly degrading protein feed, rapeseed meal. There is a 33g surplus of MP and a small surplus of ME so that liveweight gains in excess of 200g/d ought to be achieved with this diet.

Diets for fibre producing goats

The net energy requirements for fibre growth in Cashmere and Angora goats as defined by AFRC (1994) (see Chapter Two) are quite small at 0.08 and 0.25MJ/d respectively. No efficiency factor was given by ARC (1980) for fibre or wool growth so the efficiency for growth (k_f) is used here. For typical efficiencies of 0.5, daily ME requirements for fibre growth are only 0.16 and 0.5MJ, which should be added to the ME requirements in Table 8.9.

The MP requirements of fibre producing goats, however, are quite significant in amount, 13.6 and 38.5g/d for Cashmere and Angora goats respectively (see Chapter Three, p38). This is due to the low efficiency of utilisation of MP for fibre and fleece growth (k_{nw}), 0.26, so that the MP requirements of Angora goats are increased by 14 and 40gMP/d respectively above the MP requirements given in Table 8.9. As a consequence, this class of goat requires a higher protein level in its diet, as shown by Shahjalal *et al.* (1992), and as the next example shows:

Example adult Angora female goat diet 7:

Assumptions: Forages are barley straw and grass silage fed to a *housed*, female 30kg Angora goat, gaining 100g/d.
MER = 9.5MJ, L = 1.6, MP = 95g, r = 0.05

Feed name	Fresh wt,kg	DM kg	ME MJ	FME MJ	CP g	ERDP g	DUP g	MP g
Grass silage	2.4	0.60	6.6	4.9	85	59	9	
Dried sugarbeet pulp	0.1	0.10	1.3	1.2	10	5	4	
Soyabean meal	0.2	0.15	2.0	1.9	75	47	22	
Totals	2.7	0.85	9.8	**8.0**	170	111	35	**84**
Requirements		0.89	9.5	11.7		77	37	**95**

Diet calculation checks: ME intake satisfactory and is within DM appetite
M/D = 11.6MJ/kgDM, CP = 200g/kgDM
y   = 9.6g/MJ FME (from L = 1.6)
FME is limiting MCP supply, so
MPS = 77 x 0.6375 + 35 = **84g/d**
MPR = **95g/d**, a deficit of 11gMP/d

The inclusion of soyabean meal supplies over 20g/d of DUP, which balances the low [DUP] content of the grass silage. Rapeseed meal would not meet this need, due to its high degradability, but would be adequate for Cashmere goats, which require 30gMP/d less. The dietary [CP] of 200g/kg is due to the excess ERDP, 36g/d, coming from the grass silage and soyabean meal.

References

ADAS, 1991. *Technical Bulletin 91/5,* Fermentable Metabolisable Energy Content of Grass Silages, ADAS Feed Evaluation Unit, Stratford-upon-Avon.

ADAS, 1994a. *Technical Bulletin 94/5,* Equations for Predicting the Digestibility and Metabolisable Energy Concentrations of Forages, ADAS Feed Evaluation Unit, Stratford-upon-Avon.

ADAS, 1994b. *Technical Bulletin 94/9,* Rapeseed Meal, ADAS Feed Evaluation Unit, Stratford-upon-Avon.

ADAS/DANI/SAC/UKASTA, 1993. The NIRS regression equations for silage evaluation. Press Release, UKASTA, London, May 1993.

AFRC, 1990. Technical Committee on Responses to Nutrients, Report No.5. Nutritive Requirements of Ruminant Animals: Energy. *Nutr. Abs & Rev., Series B,* 60, (10), 729-804, CAB International, Wallingford, Oxon.

AFRC, 1991a. Technical Committee on Responses to Nutrients, Report No.6. A Reappraisal of the Calcium and Phosphorus Requirements of Sheep and Cattle. *Nutr. Abs & Rev., Series B,* 61, (9), 573-612, CAB International, Wallingford, Oxon.

AFRC, 1991b. Technical Committee on Responses to Nutrients, Report No.8. Voluntary Intake of Cattle. *Nutr. Abs & Rev., Series B,* 61, (11), 815-823, CAB International, Wallingford, Oxon.

AFRC, 1992. Technical Committee on Responses to Nutrients, Report No.9. Nutritive Requirements of Ruminant Animals: Protein. *Nutr. Abs & Rev., Series B,* 62, (12), 787-835, CAB International, Wallingford, Oxon.

AFRC, 1994. Technical Committee on Responses to Nutrients, Report No.10. Nutrition of Goats. *Nutr. Abs & Rev., Series B, In press,* CAB International, Wallingford, Oxon.

Alderman, G., 1982. Provisional nutrient standards for goats, ADAS Science Service, Nutrition Chemistry internal technical memorandum.

Alderman, G., 1987. Comparison of rations calculated in the different systems. In: *Feed evaluation and protein requirement systems for ruminants.* Eds. Jarrige, R., & Alderman, G., CEC, Luxembourg, pp.283-297.

Allen, D.M., 1990. *Planned Beef Production and Marketing,* BSP Professional Books, Blackwells, London.

Allen, D.M., 1992. *Rationing Beef Cattle,* Chalcombe Publications, Canterbury.

ARC, 1965. *The Nutrient Requirements of Farm Livestock,* No.2 Ruminants. Technical Review by an Agricultural Research Council Working Party, HMSO.

ARC, 1980. *The Nutrient Requirements of Ruminant Livestock.* Technical Review by an Agricultural Research Council Working Party, Commonwealth Agricultural Bureau, Farnham Royal, UK.

ARC, 1984. *The Nutrient Requirements of Ruminant Livestock,* Supplement No.1. Report of the Protein Group of the ARC Working Party, Commonwealth Agricultural Bureau, Farnham Royal, UK.

Badamana, M.S., Sutton, J.D., Oldham, J.D., & Mowlem, A., 1990. The effect of amount of protein in the concentrates on hay intake and rate of passage, diet digestibility and milk production in British Saanen goats. *Anim.Prod.,* 51, 333.

Baker, C.W., & Barnes, R.J., 1990. The application of NIRS to forage evaluation in ADAS. In: *Feedstuffs Evaluation,* Eds. Wiseman, J., & Cole, D.J.A., pp.337-352, Butterworths, London.

Barber, G.D., Givens, D.I., Kridis, M.S., Offer, N.W., & Murray, I., 1990. Prediction of the organic matter digestibility of grass silage. *Anim.Feed Sci.Technol.,* 28, 115-128.

Barber, W.P.B., Adamson, A.H., & Altman, J.F.B., 1984. New methods of feed evaluation. In: *Recent Advances in Animal Nutrition - 1984,* Eds. Haresign, W., & Cole, D.J.A., pp.161-176, Butterworths, London.

Barnes, R.J., Dhanoa, M.S., & Lister, S.J., 1989. Standard normal variate transformation and de-trending of near infra red diffuse reflectance spectra. *Appl.Spectrosc.,* 43, 772-777.

Beever, D.E., Gill, M., Dawson, J.M., & Buttery, P.J., 1990. The effect of fishmeal on the digestion of grass silage by growing cattle. *Brit.J.Nutr.,* 63, 489.

Blaxter, K.L., 1976. The Feeding of Dairy Cows for Optimal Production. George Scott Robertson Memorial Lecture, Queens University, Belfast. Rowett Research Institute Reprint No.78.

Blaxter, K.L., & Boyne, A.W., 1970. A new method of expressing the nutritive value of feeds as sources of energy. In: *Energy Metabolism of Farm Animals,* 5th Symp., Vitznau, Eds. Schurch, A., & Wenk, C., pp.9-13, Juris Druck, Zurich. (Pub. Eur. Ass. Anim. Prod. 13.

Blaxter, K.L., & Clapperton, J.L., 1965. Prediction of the amount of methane produced by ruminants. *Brit.J.Nutr.,* 19, 511.

Brett, D.J., Corbett, J.L., & Inskip, M.W., 1972. Estimation of the energy value of ewe milk. *Proc.Aust.Anim.Prod.,* 9, 286.

CEC, 1987. *Feed evaluation and protein requirement systems for ruminants.* Eds. Jarrige, R., & Alderman, G., Office for Official Publications of the European Communities, Luxembourg.

Clancy, M.J., & Wilson, R.K., 1966. *Proc.10th.Int.Grassld.Cong,* Helsinki, p.445.

Cockburn, J.E., Dhanoa, M.S., France, J., & Lopez, S., 1993. Overestimation of solubility when using dacron bag methodology. *Anim.Prod.* 55, Abst.188, *in press.*

Curran, M.K., Wimble, R.H., & Holmes, W., 1970. Prediction of the voluntary intake of food by dairy cows. *Anim.Prod.,* 12, 198-212.

CVB, 1991. *Veevoedertabel,* Centraal Veevoederbureau in Nederland, Lelystad.

Dawson, J.M., Bruce, C.J., Buttery, P.J., Gill, M., & Beever, D.E., 1988. Protein metabolism in the rumen of silage-fed steers; Effect of fishmeal supplementation. *Brit.J.Nutr.,* 60, 339-353.

Donald, H.P., & Russell, W.S., 1970. The relationship between liveweight of the ewe at mating and weight of newborn lamb. *Anim.Prod.,* 12, 273-280.

Dowman, M.G., & Collins, F.C., 1982. The use of enzymes to predict the digestibility of animal feeds. *J.Sci.Fd.Agric.,* 33, 689.

Dunshea, F.R., Bell, A.W., & Trigg, T.E., 1990. Body composition changes in goats during early lactation estimated using a two-pool model of tritiated water kinetics. *Brit.J.Nutr.*, 64, 121-131.

Gibb, M.J., & Baker, R.D., 1992. The use of fishmeal and monensin as supplements to grass silage and their effects on body composition changes in steers from 5 months of age to slaughter. *Anim.Prod.*, 55, 47-57.

Gibb, M.J., Ivings, W.E., Dhanoa, M.S., & Sutton, J.D., 1992. Changes in body components of autumn-calving Holstein-Friesian cows over the first 29 weeks of lactation. *Anim.Prod.*, 55, 339-360.

Gill, M., Beever, D.E., Buttery, P.J., England, P., Gibb, M.J., & Baker, R.D., 1987. The effect of oestradiol-17ß implantation on the response in voluntary intake, liveweight gain and body composition, to fishmeal supplementation and silage offered to growing calves. *J.Agric.Sci.*, 108, 9-16.

Givens, D.I., 1989. Predicting the Metabolisable Energy content of high temperature dried grass and lucerne. In: *Proceedings of Dri-Crop 89, 4th Int.Green Crop Drying Congress.*, Ed. Raymond, W.F., pp.117-121, British Association of Green Crop Driers.

Givens, D.I., 1990. The digestibility of maize silage - an update. *Proc. 2nd Annual Conference of the Maize Growers Association.*

Givens, D.I., Adamson, A.H., & Cobby., J.M., 1988. The effect of ammoniation on the nutritive value of wheat, barley and oat straws. II. Digestibility and energy value measurements *in vivo* and their prediction from laboratory measurements. *Anim.Feed Sci.Technol.*, 19, 173-184.

Givens, D.I., Everington, J.M., & Adamson, A.H., 1989. The digestibility and ME content of grass silage and their prediction from laboratory measurements. *Anim.Feed Sci.Technol.*, 24, 27-43.

Givens, D.I., Everington, J.M., & Adamson, A.H., 1990. The nutritive value of Spring-grown herbage produced on farms throughout England & Wales over 4 years. III. The prediction of energy values from various laboratory measurements. *Anim.Feed Sci.Technol.*, 27, 185-196.

Givens, D.I., Moss, A.R., & Adamson, A.H., 1992. The chemical composition and energy value of high temperature dried grass produced in England. *Anim.Feed Sci.Technol.*, 36, 215-218.

Goering, H.K., & Van Soest, P.J., 1970. *Forage Fiber Analysis*, Agr.Handbook No.379. Agr.Res.Serv., USDA, Washington, D.C.

Harkins, J., Edwards, R.A., & McDonald, P., 1974. A new Net Energy system for ruminants. *Anim.Prod.*, 19, 141-148.

Hulme, D.J., Kellaway, R.C., & Booth, P.J., 1986. The CAMDAIRY Model for formulating and analysing dairy cow rations. *Agric.Syst.*, 22, 81-108.

Hvelplund, T., & Madsen, J., 1990. *A study of the quantitative nitrogen metabolism in the gastro-intestinal tract, and the resultant new protein evaluation system for ruminants. The AAT-PBV system.* Thesis, Institute of Animal Science, The Royal Veterinary and Agricultural University, Copenhagen.

INRA, 1978. *Alimentation des Ruminants,* Ed. INRA Publications, Versailles.

INRA, 1988. *Alimentation des bovins, ovins, et caprins.* Ed. Jarrige, R., INRA, Paris.

Lewis, M., 1981. Equations for predicting silage intake by beef and dairy cattle. *Proc. 6th Silage Conference,* Edinburgh.

Madsen, J., 1985. The basis for the Nordic protein evaluation system for ruminants. The AAT/PBV system. *Acta Agric.Scand.*, Suppl.25, 9-20.

MAFF, 1976. *Energy Allowances and Feeding Systems for Ruminants,* MAFF Technical Bulletin 33, HMSO, London.

MAFF, 1984. *Energy Allowances and Feeding Systems for Ruminants,* ADAS Reference Book 433, HMSO, London.

MAFF, 1989. *Tables of Rumen Degradability Values for Ruminant Feedstuffs,* ADAS Feed Evaluation Unit, Stratford-upon-Avon, Warwickshire.

MAFF, 1990. *UK Tables of Nutritive Value and Chemical Compostion of Feedingstuffs,* Ministry of Agriculture, Fisheries & Food Standing Committee on Tables of Feed Composition, Eds. Givens, D.I., & Moss, A.R., Rowett Research Services Ltd, Aberdeen.

MAFF, 1991. *The Feeding Stuffs Regulations 1991,* Statutory Instrument 1991 No. 2840, HMSO, London.

MAFF, 1992. *Feed Composition - UK Tables of Feed Composition and Nutritive Value for Ruminants,* 2nd Edition, Ministry of Agriculture, Fisheries & Food Standing Committee on Tables of Feed Composition, Chalcombe Publications, Canterbury.

MAFF, 1993. *Prediction of Energy Value of Compound Feedingstuffs for Farm Animals,* Booklet 1285, MAFF Publications, Alnwick.

Mayne, C.S., & Gordon, F.J., 1985. The effect of concentrate-to-forage ratio on the milk yield response to supplementary protein. *Anim.Prod.,* 41, 269-279.

Miller, E.L., & Pike, I.H., 1987. Feeding for Profitable Beef Production. Bulletin, Association of Fish Meal Manufacturers, Potters Bar.

MLC, 1988. *Feeding the Ewe,* 3rd Edition, Meat & Livestock Commission, Milton Keynes.

MMB, 1992. *UK Dairy Facts & Figures 1992 Edition,* Federation of Milk Marketing Boards, Thames Ditton.

Morant, S.V., & Gnanasakthy, A., 1989. A new approach to the formulation of lactation curves. *Anim.Prod.,* 49, 151-162.

Moss, A.R., & Givens, D.I., 1990. Chemical composition and *in vitro* digestion to predict digestibility of field-cured and barn-dried grass hays. *Anim.Feed Sci.Technol.,* 31, 125-138.

Moss, A.R., Givens, D.I., & Everington, J.M., 1990. The effect of sodium hydroxide treatment on the chemical composition, digestibility and digestible energy content of wheat, barley and oat straws. *Anim.Feed Sci.Technol.,* 29, 73-87.

Mowlem, A., 1988. *Goat Farming.* Farming Press Books, Ipswich.

Neal, H.D.St.C., France, J., Orr, R.J., & Treacher, T.T., 1985. A model to maximise hay intake when formulating rations for pregnant ewes. *Anim.Prod.,* 40, 93-100.

NRC, 1985. *Ruminant Nitrogen Usage.* US National Academy of Science, Washington.

Oldham, J.D., 1984. Protein-energy relationships in dairy cows. *J.Dairy Sci.,* 67, 316.

Oldham, J.D., 1987. Testing and implementing the modern systems: UK. In: *Feed Evaluation and Protein Requirement Systems for Ruminants,* Eds. Jarrige, R., & Alderman, G., CEC, Luxembourg, pp.171-186.

Ørskov, E.R., & Mehrez, A.Z., 1977. Estimation of extent of protein degradation from basal feeds in the rumen of sheep. *Proc.Nut.Soc.,* 36, 78A.

Ørskov, E.R., & McDonald, I., 1979. The estimation of protein degradability in the rumen from incubation measurements weighted according to rate of passage. *J.Agric.Sci.,* 92, 499-503.

Phipps, R.H., & Wilkinson, J.M., 1985. *Maize Silage,* Chalcombe Publications, Canterbury.

Robinson, J.J., McDonald, I., Fraser, C., & Crofts, R.M.J., 1977. Studies on reproduction in prolific ewes. I. Growth of the products of conception. *J.Agric.Sci.,* 88, 539-552.

Rowlands, G.J., Lucey, S., & Russell, A.M., 1982. A comparison of different models of lactation curves in dairy cattle. *Anim.Prod.,* 35, (1), 135-144.

Roy, J.H.B., 1980. *The Calf,* 4th Edition, pp.29-65, Butterworths, London.

Russel, A.J.F., Peart, J.N., Eadie, J., MacDonald, A.J., and White, I.R., 1979. The effect of energy intake during late pregnancy on the production from two genotypes of suckler cow. *Anim.Prod.,* 28, 309-327.

Sanderson, R., Thomas, C., & McAllan, A.B., 1992. Fishmeal supplementation of grass silage given to young, growing steers: effect on intake, apparent digestibility and live-weight gains. *Anim.Prod.,* 55, 389-396.

Šebek, J.B.J., & Everts, H., 1992. Prediction of the gross energy of ewe milk. *Anim.Prod.,* 56, 101-106.

Shahjalal, Md., Galbraith, H., & Topps, J.H., 1992. The effect of changes in dietary protein and energy on growth, body composition and mohair fibre characteristics of British Angora goats. *Anim.Prod.,* 54, 405-412.

Somerville, S.H., Loman, B.G., & Edwards, R.A., 1983. A study of the relationship between plane of nutrition during lactation and certain production characteristics in autumn-calving suckler cows. *Anim.Prod.,* 37, 353-363.

Sutton, J.D., & Mowlem, A., 1991. Milk production by dairy goats. *Outlook on Agriculture,* 20, (1), 45-49.

Thomas, C., Gill, M., & Austin, A.R., 1980. The effects of supplements of fishmeal and lactic acid on voluntary intake of silage by calves. *Grass and Forage Science,* 35, 275-279.

Thomas, P.C., Robertson, S., Chamberlain, D.G., Livingstone, R.M., Garthwaite, P.H., Dewey, P.J.S., & Cole, D.J.A., 1988. Predicting the Metabolisable Energy content of compounded feeds for ruminants. In: *Recent Advances in Animal Nutrition - 1988,* Eds. Haresign, W., & Cole, D.J.A., pp.127-146, Butterworths, London.

Tilley, J.M.A., & Terry, R.A., 1963. A two stage technique for *in vitro* digestion of forage crops. *J.Br.Grassld.Soc.,* 18, 104-111.

Tyrell, H.F., & Reid, J.T., 1965. Prediction of the energy value of cow's milk. *J.Dairy Sci.,* 48, 1215-1223.

Van Straalen, W.M., & Tamminga, S., 1990. Protein degradation of ruminant diets. In: *Feedstuffs Evaluation,* Eds. Wiseman, J., & Cole, D.J.A., pp.55-72, Butterworths, London.

Vladiveloo, J., & Holmes, W. 1979. The prediction of the voluntary feed intake of dairy cows. *J.Agric.Sci.,* 93, 553-562.

Waters, C.J., Kitcherside, M.A., & Webster, A.J.F., 1992. *Anim.Feed Sci.Technol.,* 39, 279-291.

Webster, A.J.F., 1987. Metabolisable Protein - The UK Approach. In: *Feed evaluation and protein requirement systems for ruminants,* Eds. Jarrige, R., & Alderman, G., pp.47-52 CEC, Luxembourg.

Webster, A.J.F., 1992. The Metabolisable Protein System for Ruminants. In: *Recent Advances in Animal Nutrition - 1992,* Eds. Garnsworthy, P.C., Haresign, W., & Cole, D.J.A., pp.93-110, Butterworths, London.

Webster, A.J.F., Kitcherside, M.A., Kierby, J.R., & Hall, P.A., 1984. Evaluation of protein feeds for dairy cows. *Anim.Prod.,* 38, 548, Abst.

Wilkinson, J.M., 1984. *Milk and Meat from Grass,* Granada, London.

Wilkinson, J.M., & Stark, B.A., 1987. The nutrition of goats. In: *Recent Advances in Animal Nutrition - 1987,* Eds. Haresign, W., & Cole, D.J.A., pp.91-106, Butterworths, London.

Williams, R.J., 1993a. An empirical model for the lactation curve of white British dairy goats. *Anim.Prod.*, <u>57</u>, 91-97.

Williams, R.J., 1993b. Influence of farm, parity, season and litter size on the lactation curve parameters of white British dairy goats. *Anim.Prod.*, <u>57</u>, 99-104.

Wood, P.D.P., 1967. Algebraic model of the lactation curve in cattle. *Nature,* London, <u>216</u>, 164-165.

Wood, P.D.P., 1976. Algebraic models of lactation curves for milk, fat and protein production, with estimates of seasonal variation. *Anim.Prod.*, <u>22</u>, 35-40.

Wright, I.A., & Russel, A.J.F., 1987. The effect of sward height on beef cow performance and on the relationship between calf milk intake and herbage intakes. *Anim.Prod.*, <u>44</u>, 363-370.

Wright, I.A., Whyte, T.K., & Osoro, K., 1990. The herbage intake and performance of autumn-calving beef cows and their calves when grazing continuously at two sward heights. *Anim.Prod.*, <u>51</u>, 85-92.

Wright, I.A., 1992. The response of spring-born suckled calves to the provision of supplementary feeding when grazing two sward heights in autumn. *Anim.Prod.*, <u>54</u>, 197-202.

Appendix I

Tables of Feed Composition

Sources of feed data used

Whilst good progress has been made in the last 20 years in the compilation and publication of recent and validated data on the composition of ruminant feeds under the guidance of the MAFF Standing Committee on Tables of Feed Composition (MAFF 1990;1992), the situation with regard to the protein degradabilities of ruminant feeds in the UK is much less satisfactory, with adequate data on only 30 feeds published in MAFF (1989), plus a few extra feeds in MAFF (1990). The UK Metabolisable Protein is unique in requiring not only the Ørskov & McDonald (1979) fitted parameters *a, b* and *c* for the N degradability of the protein (dg) but also the [ADIN] content of the feed for the calculation of the digestibility of the Undegraded Protein, [DUP]/[UDP] or *dup*.

Two other published protein systems, INRA (1988) and the Nordic (Madsen 1985) use the Ørskov & Mehrez (1977) *in situ* dacron bag procedure to calculate the feed protein parameters used in their system, but quote degradabilities calculated for specified outflow rates as suggested by Ørskov & McDonald (1979), 6% for INRA and 8% for the Nordic system. The INRA (1988) tables are very extensive, but N degradabilities (dg6), are only mentioned in Table 13.3, where mean values for 77 feeds and forages are quoted. More importantly, this Table also lists mean values for the true digestibility of UDP of the feed in the small intestine, assigned the symbol *dsi*. Considerable use has been made of the INRA (1988) *dsi* data in compiling the tables that follow in this Manual.

Hvelplund & Madsen (1990) list the N degradabilities, at an outflow rate of 0.08 (dg8), for about 200 feeds and forages, plus buffer solubility measurements on the raw materials used in compound feed formulation (*Sol* in this Manual's tables). Use has been made of the buffer solubility values, which are quoted where they are available, as a guide to the estimation of the *a* fraction (cold water extractable) of feeds.

Van Straalen & Tamminga (1990) give two tables on the N degradabilities of raw materials. Table 4.2 gives the *a* and *c* fractions for 28 raw materials, plus the undigestible N fraction (*u*), from which *b* can be calculated, since:

$$a + b + u = 1 \tag{148}$$

as discussed in Chapter Four earlier. These authors also give the protein bypass fraction as percent at an outflow rate of 0.06, from which the degradability (dg6), as a decimal fraction is easily calculated. Their Table 4.3 gives the mean bypass protein values for 50 raw materials, corrected for inter-laboratory variation. These are high quality data, easily adapted to the calculation of the parameters [ERDP] and [DUP] at other outflow rates as required by the MP system. The Nordic and French degradability data are less easily used, but are useful as a cross check on the soundness of other data.

Whilst the UK animal feed industry has extensive and confidential individual company databases on raw material degradabilities, relatively few of these data have been released into the public domain. However, the UKASTA Scientific Committee has recently released some additional data on less common raw materials, and also agreed typical mean degradability parameters and [ADIN] contents for the range of ruminant compound feeds on the market in the UK.

Principles used in compiling the tables

Since many of the cereals, legumes, byproducts, oilseed meals and other raw materials used in compound feed formulation and ruminant feeding in the UK are traded internationally, it seemed reasonable to make use of the Dutch, French and Nordic data on these feeds in compiling the tables that follow. Little use has been made of the extensive European data on fresh forages and silages, since forage varieties and conservation practices vary considerably between countries, and the MAFF database gave fairly good coverage of forages, except for cereal silages and root crops. The Nordic and French data were used for these forages.

The MAFF (1990;1992), ADAS *et al.* (1993) and (Van Straalen & Tamminga 1990) data for the parameters *a, b* and *c* were used to compute the range of [ERDP] values required, using equations (27), (28) and (29) in Chapter One. Degradabilities at an outflow rate of 0.08 were calculated using equation (26) and tabulated for comparison with other data. The Nordic data are for an outflow rate, r = 0.08, so comparisons with MAFF and UKASTA data are possible and are quoted on the degradability parameter pages of the tables.

Where [ADIN] values were available, ie MAFF (1990;1992) and ADAS *et al.* (1993), these have been used to calculate [DUP] values in accordance with equations (32) and (33) in Chapter One. In the case of brewing and distillery by-products, which have very high [ADIN] values, the values were discounted by one half as suggested by Webster (1992). Where negative [DUP] values were obtained by using equation (33), an appropriate INRA [UDP] digestibility value *(dsi)* was used instead. The digestibility of UDP *(dup)* was also calculated for these data, for comparison with the INRA *dsi* estimates. In the case of the Dutch (NL) data, the *dsi* values of INRA (1988) were used in order to calculate [DUP] for [UDP] calculated from [CP] and *a, b* and *c* values for a feed, where these were available. In a number of cases, values for the digestibility of [UDP] were obtained from both sources, with satisfyingly good agreement.

For a limited number of feeds and forages, mainly roots, cereal silages and straws, the only sources of degradability data were INRA (1988) or Madsen (1985), yielding either dg6 or dg8 values plus *dsi* values, but no *a, b* and *c* values. To obtain estimates of [ERDP] and [DUP] values at outflow rates of 0.02, 0.05 and 0.08, the approximations outlined in Chapter Four (equations (147) and (148)) were used to obtain values for the *a* and *b* parameters. An appropriate value for *c* was then calculated using equation (149) and the relevant value for outflow rate (r). The calculation of [ERDP] and [DUP] values at other outflow rates was then possible, but the values are only estimates and are quoted in *italics* in the tables to indicate their method of calculation.

Layout of tables

The tables that follow are in two parts, 1a and 1b etc, on facing pages. The left hand page gives the data required to carry out diet formulation in accordance with the procedures laid down in this Manual and used in all examples, namely [DM], [ME], [FME], [EE], [CP], plus [ERDP] and [DUP] values for outflow rates of 0.02, 0.05 and 0.08. The value for the digestibility of UDP (*dup*) has been listed, either by calculation from [DUP]/[UDP] for MAFF and UKASTA data, or as INRA (1988) *dsi* values (marked *). Because of space limitations on headings, additional abbreviations have been introduced, which it is hoped are easily understood. Full details are given below.

The second part or right hand page gives the basal data on feed N degradability, the *a, b* and *c* values, plus the calculated undegradable fraction *u* (or 1-*a*-*b*). [ADIN] and/or INRA (1988) *dsi* values are quoted where they are available. These data are intended for use in computer programmes generating [ERDP] and [DUP] values at the exact outflow rate indicated by the level of feeding of the diet being formulated, using equation (25). The buffer solubility values from Madsen (1985) are listed where available. Finally, degradability values from all four sources are quoted alongside, Dutch (NL) and French (F) values as dg6, Nordic (DK) as dg8, plus dg8 values from MAFF or UKASTA.

Regrettably, space would not permit the inclusion of the INFIC International feed numbers used in MAFF (1990;1992), nor was it thought helpful to use the feed groupings of INFIC. Feeds are grouped by their general characteristics, origins and processes, so that users can see the range of degradabilities found in a particular feed grouping such as cereals or silages.

Index to table headings and codes

Proceeding from left to right the headings are as follows:

Left hand pages

No	Feed number to ensure that tables are read across correctly
Description	Brief feed name, annotated by data source origin, see below
DM	Dry matter content (g/kg)
ME	ME as (MJ/kgDM)
FME	Fermentable ME (MJ/kgDM)
EE	Ether extract (oil) content (g/kgDM)
CP	Crude Protein content (g/kgDM)
RP2	[ERDP] (g/kgDM) for r = 0.02
UP2	[DUP] (g/kgDM) for r = 0.02
RP5	[ERDP] (g/kgDM) for r = 0.05
UP5	[DUP] (g/kgDM) for r = 0.05
RP8	[ERDP] (g/kgDM) for r = 0.08
UP8	[DUP] (g/kgDM) for r = 0.08
dup	Digestibility of [UDP], calculated, or INRA *dsi* value marked *

Right hand pages

Code	Country code to indicate data source as follows:
UK	MAFF (1990;1992) tables
UK[1]	UKASTA data
NL	Van Straalen & Tamminga (1990), feeds annotated [2] on left page
DK	Madsen (1985) tables, feeds annotated [3]
F	INRA (1988) Table 13.3, feeds annotated [4]
CP	Crude Protein (g/kgDM)
ADIN	Acid Detergent Insoluble N (g/kgDM)
Sol	Buffer solubility of feed as decimal of total N
a	Cold water extractable N as decimal of total N (by dacron bag)
b	Slowly degradable N as decimal of total N
c	Rate of change constant for *b* fraction
u	Undegradable N as decimal of total N (= 1-*a*-*b*)
dsi	Digestibility in small intestine of UDP as decimal, INRA (1988)
dg6	Degradability of total N as decimal for outflow of 0.06/h, NL/F
dg8[3]	Degradability of total N as decimal for outflow of 0.08/h, DK
dg8	Degradability of total N as decimal for outflow of 0.08/h, UK

Figures in italics are derived by back calculation from dg6 data from either INRA (1988) or Madsen (1985), having estimated both a and b fractions as indicated in Chapter Four.

Feed Composition Table 1a - diet formulation parameters

No DESCRIPTION   *Units are g/kg dry matter except ME and FME as MJ/kgDM*

No	DESCRIPTION	DM	ME	FME	EE	CP	RP2	UP2	RP5	UP5	RP8	UP8	dup
	FRESH FORAGES												
1	Grass 55-60D	200	7.5	6.9	16	97	64	23	52	33	45	40	.82
2	Grass 60-65D	200	9.5	8.8	19	120	85	22	72	33	64	40	.82
3	Grass 65-70D	200	10.7	10.0	21	135	100	16	87	28	79	36	.76
4	Grass 70-75D	200	11.6	10.7	25	150	118	17	103	30	92	40	.79
5	Grass 75-80D	200	12.3	11.4	25	190	157	15	141	29	129	40	.76
6	Grass >80D	200	12.6	11.7	25	190	159	20	134	42	117	58	.83
7	Italian ryegrass	200	11.4	10.6	22	128	102	8	93	16	86	22	.69
	GRASS AND LEGUME SILAGES												
8	Grass 65D	250	10.3	7.6	42	140	102	11	97	15	94	18	.64
9	Grass 70D	250	11.7	8.6	47	174	125	17	120	22	116	25	.70
10	Grass UK mean	250	11.0	8.1	45	142	104	11	99	15	95	19	.65
11	Grass and clover	229	9.8	7.0	36	174	130	14	118	25	109	33	.65
12	Baled grass 55D	390	8.8	7.2	28	151	102	29	84	45	76	52	.85
13	Baled grass 65D	360	10.6	8.6	32	160	100	35	95	39	91	42	.84
14	Baled grass 75D	230	12.2	8.8	37	180	136	14	130	19	126	23	.80
15	Lucerne	338	8.0	6.4	25	194	146	10	140	16	135	20	.60
	CEREAL SILAGES												
16	Barley, whole[3]	394	9.1	7.5	20	90	*64*	*11*	*59*	*14*	*56*	*16*	.70*
17	Barley + urea	550	10.0	9.3	20	244	*176*	*20*	*171*	*24*	*168*	*26*	.70*
18	Maize, whole	295	11.3	9.0	32	98	*69*	*11*	*67*	*13*	*65*	*14*	.70*
19	Wheat, whole[3]	410	10.0	8.3	20	99	*71*	*11*	*66*	*15*	*62*	*17*	.70*
20	Wheat + urea	550	10.3	9.6	20	244	*176*	*20*	*171*	*24*	*168*	*26*	.70*
	ROOTS												
21	Carrots[3]	100	12.0	11.8	5	105	*85*	*8*	*82*	*10*	*79*	*13*	.60*
22	Fodder beet[3]	183	11.9	11.8	3	63	*52*	*5*	*49*	*7*	*47*	*8*	.65*
23	Kale[3]	140	11.8	11.1	20	160	*129*	*15*	*120*	*21*	*112*	*26*	.65*
24	Potatoes[3]	204	13.4	13.3	2	108	*89*	*8*	*85*	*11*	*81*	*13*	.60*
25	Turnips[3]	90	13.0	12.9	4	112	*92*	*9*	*88*	*12*	*84*	*15*	.65*
	DRIED FORAGES												
26	Grass hay	850	9.2	8.6	16	81	53	15	44	23	39	28	.73
27	Lucerne hay	865	8.5	8.1	13	183	141	26	131	34	123	40	.75*
28	Dried grass	917	10.7	9.4	37	199	144	23	117	48	103	60	.74
29	Alfalfa meal[2]	900	8.8	7.8	28	160	95	40	76	53	67	60	.70*
30	Dried lucerne	895	8.8	7.8	28	199	140	22	123	37	114	45	.72
	CEREAL STRAWS												
31	Barley[3]	850	6.4	5.9	14	42	*28*	*5*	*25*	*7*	*23*	*9*	.57
32	Barley+ammonia[3]	850	7.7	7.2	15	70	*50*	*3*	*48*	*5*	*46*	*6*	.46
33	Oat	850	7.2	6.7	14	34	23	5	20	7	18	9	.65
34	Oat + ammonia	850	8.0	7.4	18	78	31	35	24	42	22	44	.84
35	Wheat[3]	850	6.1	5.7	12	39	*26*	*5*	*23*	*8*	*21*	*10*	.61
36	Wheat+ammonia	850	7.3	6.9	13	68	48	1	47	2	46	3	.21
37	Wheat + NaOH	850	8.6	8.3	9	36	*26*	*5*	*24*	*6*	*23*	*6*	.70

Feed Composition Table 1b - N degradability parameters

No   **DESCRIPTION**   *Units are decimal proportions except CP and ADIN as g/kgDM*

No	Description	Code	CP	ADIN	Sol	a	b	c	u	dsi	dg6	dg8[3]	dg8
	FRESH FORAGES												
1	Grass 55-60D	UK	97	0.7	-	.16	.57	.08	.17	.75	-	-	.50
2	Grass 60-65D	UK	120	0.7	-	.28	.59	.09	.13	.75	-	.67	.59
3	Grass 65-70D	UK	135	1.2	-	.34	.57	.10	.09	.75	.73	.68	.65
4	Grass 70-75D	UK	150	1.0	-	.25	.68	.12	.07	.75	.73	.73	.66
5	Grass 75-80D	UK	190	1.3	-	.23	.71	.18	.06	.75	.73	-	.72
6	Grass >80D	UK	190	0.9	-	.08	.89	.13	.03	.75	-	-	.63
7	Ital.ryegrass	UK	128	1.2	-	.36	.57	.17	.07	.75	-	.77	.75
	GRASS AND LEGUME SILAGES												
8	Grass 65D	UK	140	1.3	-	.63	.26	.14	.11	.60	.78	.70	.80
9	Grass 70D	UK	174	1.3	-	.62	.25	.18	.13	.60	.78	.72	.79
10	Grass UK mean	UK	142	1.3	-	.63	.26	.14	.11	.60	.75	.71	.80
11	Grass and clover	UK	174	2.2	-	.42	.48	.12	.10	.60	.78	.69	.71
12	Baled grass 55D	UK	151	0.5	-	.45	.53	.03	.02	.60	-	-	.59
13	Baled grass 65D	UK	160	0.5	-	.58	.19	.10	.23	.60	.75	-	.69
14	Baled grass 75D	UK	180	0.5	-	.71	.22	.12	.07	.60	-	-	.84
15	Lucerne	UK	194	1.8	-	.66	.25	.17	.09	.55	.78	.76	.83
	CEREAL SILAGES												
16	Barley, whole	DK/F	90	-	-	*.60*	*.30*	*.07*	*.10*	*.70*	.72	.74	.74
17	Barley + urea	Estd.	244	-	-	*.80*	*.10*	*.08*	*.10*	*.70*	-	-	*.85*
18	Maize, whole	UK	98	-	-	.66	.19	.20	.15	.70	.72	.62	.80
19	Wheat, whole	DK/F	99	-	-	*.60*	*.30*	*.08*	*.10*	*.70*	.72	.75	.75
20	Wheat + urea	Estd.	24	-	-	*.80*	*.10*	*.07*	*.03*	*.70*	-	-	*.85*
	ROOTS												
21	Carrots	DK/F	105	-	-	*.25*	*.65*	*.44*	*.10*	.60	.85	.80	.80
22	Fodder beet	DK/F	63	-	-	*.25*	*.65*	*.44*	*.10*	.65	.85	.80	.80
23	Kale	DK/F	160	-	-	*.25*	*.65*	*.27*	*.10*	.65	-	.75	.75
24	Potatoes	DK/F	108	-	-	*.25*	*.65*	*.44*	*.10*	.60	.65	.80	.80
25	Turnips	DK/F	112	-	-	*.25*	*.65*	*.34*	*.10*	.65	.85	.80	.80
	DRIED FORAGES												
26	Grass hay	UK	81	1.2	-	.22	.60	.08	.18	.70	.66	.72	.52
27	Lucerne hay	DK/F	183	-	-	*.25*	*.65*	*.29*	*.15*	*.75*	.66	.71	.71
28	Dried grass	UK	199	2.3	-	.37	.63	.04	.00	.70	.60	-	.59
29	Alfalfa meal	NL/F	160	-	-	.26	.54	.05	.20	.70	.52	-	.47
30	Dried lucerne	UK	199	2.0	-	.56	.38	.04	.06	.70	.60	-	.69
	CEREAL STRAWS												
31	Barley	DK/F	42	1.0	-	*.30*	*.50*	*.12*	*.20*	*.70*	.60	.60	-
32	Barley+ammonia	DK/F	70	1.1	-	*.70*	*.20*	*.08*	*.10*	*.70*	-	.80	-
33	Oat	UK	34	0.6	-	.30	.51	.11	.19	.70	.60	.60	-
34	Oat + ammonia	UK	78	1.0	-	.27	.54	.01	.19	.70	-	-	.33
35	Wheat	UK/DK	39	0.8	-	*.30*	*.50*	*.12*	*.20*	*.70*	.60	.60	-
36	Wheat+ammonia	UK	68	1.5	-	.74	.13	.13	.13	.70	-	.80	.82
37	Wheat + NaOH	DK/F	36	-	-	*.50*	*.35*	*.20*	*.15*	*.70*	-	.75	-

Feed Composition Table 2a - diet formulation parameters

No DESCRIPTION   *Units are g/kg dry matter except ME and FME as MJ/kgDM*

No	DESCRIPTION	DM	ME	FME	EE	CP	RP2	UP2	RP5	UP5	RP8	UP8	dup
	CEREALS												
38	Barley, ground	840	13.3	12.7	16	129	111	8	105	14	99	19	.80
39	Maize, ground[2]	860	13.8	12.4	40	102	*45*	*49*	*33*	*61*	*29*	*64*	.95*
40	Oats, ground	860	12.1	10.7	41	105	83	6	82	8	81	9	.95*
41	Oats, ground[3]	860	12.1	10.7	41	108	*91*	*8*	*86*	*13*	*82*	*16*	.95*
42	Oats, naked	860	14.8	11.7	90	128	102	3	101	5	99	6	.63
43	Triticale, ground	857	13.8	13.3	15	131	105	10	101	14	97	17	.95*
44	Wheat, ground	860	13.7	13.1	17	128	108	6	104	10	100	13	.80
45	Wheat, ground[1]	860	13.7	13.1	17	133	107	11	96	20	88	27	.80
	LEGUME SEEDS												
46	Beans	865	13.1	12.7	12	333	278	22	254	43	236	59	.86
47	Beans[1]	865	13.1	12.7	12	299	*245*	*25*	*229*	*39*	*216*	*51*	.85*
48	Lupins[2]	860	14.2	10.5	104	342	*288*	*22*	*251*	*44*	*226*	*59*	.60*
49	Peas[2]	868	13.5	13.0	14	252	*204*	*16*	*184*	*32*	*172*	*42*	.80*
	CEREAL BYPRODUCTS												
50	Maize gluten feed	890	12.9	11.4	44	207	162	10	149	22	140	29	.71
51	Maize gluten fd.[1]	890	12.9	11.4	44	194	160	12	143	29	130	40	.81
52	Maize gluten meal	905	17.5	16.4	31	666	346	242	232	345	181	391	.82
53	Maize glut. meal[1]	905	17.5	16.4	31	689	102	483	68	513	60	521	.84
54	Oat feed	860	5.6	4.7	25	51	30	7	27	9	26	10	.53
55	Rice bran[1]	886	7.1	6.9	7	178	109	45	98	55	90	62	.80
56	Rice bran[2]	886	7.1	6.9	7	142	*88*	*45*	*65*	*65*	*52*	*76*	.85*
57	Salseed	860	11.0	10.5	13	93	64	12	48	26	42	32	.72
58	Soya hulls[1]	860	13.2	12.6	18	141	102	19	80	39	67	50	.75
59	Wheat feed[1]	879	11.9	10.4	44	188	142	28	125	43	113	54	.86
60	Wheat feed[2]	879	11.9	10.4	44	184	*147*	*31*	*125*	*51*	*110*	*66*	.95*
	BREWERY BYPRODUCTS												
61	Brewers grains	250	11.5	9.3	62	218	*147*	*54*	*118*	*78*	*101*	*92*	.85*
62	Dr.brewers grns[1]	860	12.2	9.9	67	249	*126*	*103*	*91*	*132*	*72*	*148*	.85*
63	Dist.grains,maize	889	14.7	10.9	110	317	185	64	154	92	137	107	.67
64	Dist.grains,wheat	890	12.4	10.5	55	302	248	9	242	12	238	15	.50
65	Dist.grains+sols	860	13.8	9.5	122	293	170	17	129	53	110	70	.42
	PULPS												
66	Beet+molasses	880	12.5	12.4	4	103	63	25	49	37	43	43	.81
67	Beet+molasses[1]	880	12.5	12.4	4	117	90	7	78	18	71	25	.69
68	Beet unmolassed[2]	890	12.8	12.6	7	103	*71*	*19*	*56*	*30*	*48*	*35*	.70*
69	Citrus pulp	890	12.6	11.8	22	99	78	2	68	11	63	16	.60
70	Citrus pulp[2]	890	12.6	11.8	22	70	*52*	*10*	*44*	*16*	*40*	*20*	.80*
	MISCELLANEOUS FEEDS												
71	Fat prills	950	33.0	0	950	0	0	0	0	0	0	0	0
72	Feather meal[2]	900	13.0	11.7	38	889	321	*273*	207	*330*	169	*349*	.50*
73	Fishmeal	930	14.2	11.0	91	694	380	232	287	311	250	342	.85*
74	Molasses[3]	750	13.1	13.0	4	118	*94*	*0*	*94*	*0*	*94*	*0*	0
75	Urea	950	0	0	0	2600	2080	0	2080	0	2080	0	0

Feed Composition Table 2b - N degradability parameters

No	DESCRIPTION	*Units are decimal proportions except CP and ADIN as g/kgDM*									

No		Code	CP	ADIN	Sol	a	b	c	u	dsi	dg6	dg8[3]	dg8
	CEREALS												
38	Barley, ground	UK	129	0.4	.22	.25	.70	.35	.05	.85	.74	.70	.82
39	Maize, ground[2]	DK	102	-	.27	*.26*	*.69*	*.01*	*.05*	.95	.42	.31	.31
40	Oats, ground	UK	105		.52	.72	.23	.40	.05	.95	.78	.84	.91
41	Oats, ground	DK/F	108		.52	.39	.56	.33	.05	.95	.78	.84	.84
42	Oats, naked	UK	128	0.5	-	.74	.22	.36	.04	.95	-	-	.92
43	Triticale, ground	UK	131	-	-	.59	.36	.24	.05	.95	.76	-	.86
44	Wheat, ground	UK	128	0.3	.31	.45	.51	.38	.04	.95	.74	.82	.87
45	Wheat, ground	UK	133	0.6	.31	.39	.57	.13	.04	.95	.74	-	.74
	LEGUME SEEDS												
46	Beans	UK	333	0.5	.66	.42	.56	.16	.02	.60	.86	.86	.79
47	Beans	UK	299	0.5	.66	.33	.72	.09	-.05	.60	.86	.86	.71
48	Lupins	NL	342	-	.65	.26	.73	.13	.01	.60	.78	.80	.71
49	Peas	NL	252	-	.69	.56	.44	.09	.00	.80	.76	.77	.79
	CEREAL BY-PRODUCTS												
50	Maize gluten fd	UK	207	1.4	.49	.61	.36	.09	.03	.85	.69	.68	.80
51	Maize gluten fd	UK	194	0.8	.49	.38	.61	.12	.01	.85	.69		.75
52	Maize glut. meal	UK	666	6.4	.09	.08	.76	.03	.16	.90	.27	.17	.29
53	Maize glut. meal	UK	689	6.4	.09	.08	.92	.002	.00	.90	.27	.17	.10
54	Oat feed	UK	51	1.3	-	.51	.22	.07	.27	.80	-	-	.61
55	Rice bran	UK	178	2.6	-	.29	.60	.06	.11	.85	.66	-	.55
56	Rice bran	NL	142	-	.25	.04	.78	.06	.18	.85	.66	-	.37
57	Salseed	UK	93	1.4	-	.37	.78	.02	-.15	.80	-	-	.53
58	Soya hulls	UK	141	1.8	-	.23	.76	.05	.01	.70	.57	-	.52
59	Wheat feed	UK	188	0.4	.31	.34	.57	.11	.09	.95	.76	-	.67
60	Wheat feed	NL	184	-	.31	.13	.80	.13	.07	.95	.69	.82	.63
	BREWERY BYPRODUCTS												
61	Brewers grains	DK	218		.11	*.19*	*.67*	*.07*	*.14*	.85	.45	.49	.49
62	Dr.brewers grns.	NL	249	-	.11	.05	.65	.05	.30	.50	.39	.49	.49
63	Dist.grains,maize	UK	317	13.0	-	.32	.46	.05	.22	-	-	-	.50
64	Dist.grains,wht.	UK	302	13.0	-	.84	.12	.11	.04	-	-	-	.91
65	Dist.grains + sols	UK	293	14.3	-	.26	.62	.03	.12	-	-	-	.43
	PULPS												
66	Beet + molasses	UK	103	0.9	.39	.32	.58	.03	.10	.70	.61	.65	.48
67	Beet + molasses	UK	117	1.3	.39	.47	.53	.06	.00	.70	.61	.65	.70
68	Beet unmolassed	NL	103	-	.12	.24	.70	.05	.06	.70	.51	.38	.51
69	Citrus pulp	UK	99	1.4	-	.51	.49	.07	.00	.80	.66	.63	.74
70	Citrus pulp[2]	NL	70	-	-	.41	.56	.06	.03	.80	.66	.63	.65
	MISCELLANEOUS FEEDS												
71	Fat prills	UK	0	-	-	na	na	na	na	na	na	na	na
72	Feather meal	NL	889	-	-	.13	.77	.01	.10	.50	.34	-	.22
73	Fishmeal	UK	694	-	.26	.30	.63	.02	.07	.85	.45	.43	.42
74	Molasses	DK	118	-	1.0	1.0	.00	na	.00	na	1.0	1.0	1.0
75	Urea	UK	2600	-	1.0	1.0	.00	na	.00	na	1.0	1.0	1.0

Feed Composition Table 3a - diet formulation parameters

| No | DESCRIPTION | *Units are g/kg dry matter except ME and FME as MJ/kgDM* |

No		DM	ME	FME	EE	CP	RP2	UP2	RP5	UP5	RP8	UP8	dup
	OILSEED MEALS - high fibre												
76	Babasu meal[2]	860	12.6	10.1	71	202	*110*	*54*	*71*	*78*	*53*	*89*	.60*
77	Coconut meal[2]	860	12.9	9.3	102	215	*131*	*70*	*91*	*106*	*73*	*123*	.90*
78	Cottonseed meal	924	11.1	8.8	66	375	268	70	222	109	195	132	.85*
79	Cottonseed meal[1]	924	11.1	8.8	66	385	288	52	224	110	189	141	.80
80	Nigerseed meal[2]	860	11.9	9.0	82	361	*289*	*39*	*239*	*69*	*206*	*89*	.60*
81	Palmkernel meal	890	11.6	11.3	10	170	108	34	77	62	62	74	.73
82	Palmkernel meal[1]	890	11.6	11.3	10	194	143	38	116	61	101	76	.90*
83	Rapeseed meal	890	12.0	11.0	28	400	319	29	288	57	265	78	.71
84	Rapeseed meal[1]	890	12.0	11.0	28	381	312	22	276	55	248	79	.67
85	Shea nut	860	9.2	5.0	120	164	85	31	66	48	59	54	.57
86	Sunflower meal	898	9.6	8.8	23	336	276	24	249	49	229	67	.77
87	Sunflower meal[1]	898	9.6	8.8	23	370	320	15	301	32	285	46	.68
	OILSEED MEALS - low fibre												
88	Groundnut meal[2]	860	13.7	11.3	70	570	*459*	*73*	*382*	*138*	*333*	*180*	.85*
89	Linseed meal[1]	885	11.9	8.9	87	379	305	30	267	64	242	87	.80
90	Linseed meal[2]	885	11.9	8.9	87	334	*234*	*75*	*177*	*123*	*147*	*149*	.85*
91	Linseed flakes[1]	860	11.9	8.9	87	376	309	39	259	84	225	115	.82
92	Soyabean meal	905	13.3	12.7	17	497	398	70	313	146	260	193	.85
93	Soyabean meal[1]	905	13.3	12.7	17	538	453	54	376	123	323	171	.84
	COMPOUND FEEDS												
94	Dairy, 13MJ	860	13.0	10.3	77	205	153	30	135	46	123	57	.83
95	Dairy, 12.5MJ	860	12.2	10.0	64	207	156	28	138	44	125	55	.81
96	Dairy concentrate	860	12.7	9.9	80	300	225	47	196	73	175	92	.83
97	Maize balancer	860	12.7	10.2	72	285	210	44	185	67	167	83	.83
98	Calf weaner-hay	860	12.8	11.0	53	206	155	30	137	46	125	57	.84
99	Calf weaner-silage	860	12.8	11.0	52	190	148	22	133	35	122	45	.82
100	Dry cows	860	12.0	10.3	48	166	128	19	115	31	106	39	.81
101	Beef-grass silage	860	12.5	10.8	48	170	129	22	115	34	105	43	.81
102	Beef-maize silage	860	12.8	10.7	60	180	134	26	121	38	111	47	.82
103	Beef-Rosemaund	860	12.9	11.0	55	195	147	27	132	41	121	51	.83
104	Beef-cereal diet	860	11.5	9.0	72	375	286	53	251	84	227	106	.85
105	Pregnant ewe-hay	860	12.4	10.1	65	195	148	26	132	40	121	50	.85
106	Preg.ewes-silage	860	12.4	10.1	67	184	139	25	125	38	114	47	.82
107	Lactating ewes	860	12.4	10.4	57	206	158	26	143	40	132	50	.83
108	Lambs-fattening	860	12.8	11.0	52	195	145	30	128	45	117	55	.84
109	Lamb concentrate	860	11.8	9.7	60	420	316	67	279	100	252	124	.85
110	Lambs rearer	860	12.3	10.3	57	172	131	22	118	33	108	42	.81
111	Lactating goats	860	12.5	10.4	61	206	153	30	138	44	127	54	.82

Feed Composition Table 3b - N degradability parameters

No DESCRIPTION *Units are decimal proportions except CP and ADIN as g/kgDM*

No		Code	CP	ADIN	Sol	a	b	c	u	dsi	dg6	dg8[3]	dg8
	OILSEED MEALS - high fibre												
76	Babasu meal	NL	202	-	-	.03	.87	.03	.10	.60	.38	-	.27
77	Coconut meal	NL	215	-	.13	.14	.83	.03	.03	.90	.43	.37	.37
78	Cottonseed meal	UK	375	-	.35	.33	.60	.06	.07	.85	.57	.56	.59
79	Cottonseed meal	UK	385	3.2	.35	.24	.76	.05	0	.85	.57	.56	.53
80	Nigerseed meal	NL	361	-	.50	.09	.86	.11	.05	.60	.64	-	.59
81	Palmkernel meal	UK	170	3.0	.10	.18	.82	.03	.05	.90	.38	.34	.40
82	Palmkernel meal	UK	194	-	.10	.24	.70	.07	.06	.90	.38	.34	.57
83	Rapeseed meal	UK	400	3.6	.38	.32	.61	.16	.07	.80	.71	.68	.73
84	Rapeseed meal	UK	385	4.8	.38	.19	.75	.16	.06	.80	.71	.68	.69
85	Shea nut	UK	164	5.5	-	.32	.52	.02	.16	.80	-	-	.42
86	Sunflower meal	UK	336	2.0	.43	.30	.65	.17	.05	.85	.77	.73	.74
87	Sunflower meal	UK	370	2.6	.43	.24	.71	.35	.05	.85	.77	.73	.82
	OILSEED MEALS - low fibre												
88	Groundnut meal	NL	570	-	.57	.22	.77	.09	.01	.85	.69	.74	.63
89	Linseed meal	UK	379	1.9	.29	.38	.60	.10	.02	.85	.60	.60	.71
90	Linseed meal	NL	334	-	.29	.17	.79	.05	.04	.85	.60	.60	.47
91	Linseed flakes	UK	376	2.0	-	.14	.84	.11	.02	.85	-	-	.63
92	Soyabean meal	UK	497	2.2	.30	.08	.92	.08	.00	.90	.60	.60	.54
93	Soyabean meal	UK	538	2.3	.30	.10	.90	.11	.00	.90	.61	.60	.62
	COMPOUND FEEDS												
94	Dairy, 13MJ	UK[1]	205	0.9	-	.33	.56	.12	.11	-	-	-	.67
95	Dairy, 12.5MJ	UK[1]	207	1.1	-	.33	.57	.12	.10	-	-	-	.67
96	Dairy concentrate	UK[1]	300	1.3	-	.25	.64	.12	.11	-	-	-	.63
97	Maize balancer	UK[1]	285	1.2	-	.32	.57	.11	.11	-	-	-	.65
98	Calf weaner-hay	UK[1]	206	0.7	-	.33	.57	.12	.10	-	-	-	.67
99	Calf weaner-silage	UK[1]	190	0.8	-	.34	.58	.14	.08	-	-	-	.71
100	Dry cows	UK[1]	166	0.8	-	.35	.56	.14	.08	-	-	-	.71
101	Beef-grass silage	UK[1]	170	0.8	-	.35	.55	.13	.10	-	-	-	.69
102	Beef-maize silage	UK[1]	180	0.8	-	.34	.54	.14	.12	-	-	-	.68
103	Beef-Rosemaund	UK[1]	195	0.8	-	.33	.56	.14	.11	-	-	-	.69
104	Beef-cereal diet	UK[1]	375	1.2	-	.30	.61	.12	.09	-	-	-	.67
105	Pregnant ewe-hay	UK[1]	195	0.9	-	.34	.56	.13	.10	-	-	-	.69
106	Preg.ewes-silage	UK[1]	184	0.8	-	.34	.55	.14	.11	-	-	-	.69
107	Lactating ewes	UK[1]	206	0.8	-	.35	.55	.15	.10	-	-	-	.71
108	Lambs-fattening	UK[1]	195	0.7	-	.32	.57	.12	.11	-	-	-	.66
109	Lamb concentrate	UK[1]	420	1.4	-	.25	.63	.14	.12	-	-	-	.65
110	Lambs rearer	UK[1]	172	0.8	-	.35	.55	.14	.10	-	-	-	.70
111	Lactating goats	UK[1]	206	0.9	-	.34	.54	.14	.12	-	-	-	.68

KEY to annotations *1 Data from UKASTA Scientific Committee*
2 Data from Van Straalen & Tamminga (1990)
3 Data from Hvelplund & Madsen (1990)

Appendix II

Sequential List of Equations

Energy terms
$$E = M \times k \tag{1}$$

$$ME = GE - FE - UE - M_E \tag{2}$$

$$4.184 \text{ joules} = 1 \text{ calorie} \tag{3}$$

$$q_m = [ME]/[GE] \tag{4}$$

$$[FME] \ (MJ/kgDM) = [ME] - [ME_{fat}] - [ME_{ferm}] \tag{5}$$

ME efficiencies
$$k_m = 0.35q_m + 0.503 \tag{6}$$

$$k_l = 0.35q_m + 0.420 \tag{7}$$

$$k_f = 0.78q_m + 0.006 \tag{8}$$

$$k_g = 0.95k_l \tag{9}$$

$$k_c = 0.133 \tag{10}$$

$$k_t = 0.84 \tag{11}$$

Feeding level correction
$$C_L = 1 + 0.018(L - 1) \tag{12}$$

Energy retention
$$R = B(1 - e^{-kl}) - 1 \tag{13}$$

$$B = k_m/(k_m - k_f) \tag{14}$$

$$k = k_m \times \ln(k_m/k_f) \tag{15}$$

$$ME \ (MJ/d) = E/k \tag{16}$$

Requirements M_{mp} (MJ/d) = $C_L\{E_m/k_m + E_l/k_l + E_g/k_g + E_c/k_c\}$ (17)

$$M_{mp} \text{ (MJ/d)} = (F/k) \times \ln\{B/(B - R - 1)\} \qquad (18)$$

$$x = \frac{(M/D - [ME] \text{ forage})}{([ME] \text{ compound} - [ME] \text{ forage})} \qquad (19)$$

Variable Net Energy $k_{mp} = E_{mp}/M_{mp}$ (20)

$$E_{mp} \text{ (MJ/d)} = M_{mp} \times k_{mp} \qquad (21)$$

Protein terms   DMTP = 0.75 × 0.85 × MCP = 0.6375MCP   (22)

$$MP \text{ (g/d)} = 0.6375MCP + DUP \qquad (23)$$

$$dg = a + b\{1 - e^{(-ct)}\} \qquad (24)$$

$$r = -0.024 + 0.179\{1 - e^{(-0.278L)}\} \qquad (25)$$

$$p = a + (b \times c)/(c + r) \qquad (26)$$

$$[QDP] \text{ (g/kgDM)} = a \times [CP] \text{ (g/kgDM)} \qquad (27)$$

$$[SDP] \text{ (g/kgDM)} = \{(b \times c)/(c + r)\} \times [CP] \text{ (g/kgDM)} \qquad (28)$$

$$[ERDP] \text{ (g/kgDM)} = 0.8[QDP] + [SDP] \qquad (29)$$

$$ERDP \text{ (g/d)} = w_1[ERDP_1] + w_2[ERDP_2] + w_3[ERDP_3] \text{ etc} \qquad (30)$$

$$[UDP] \text{ (g/kgDM)} = [CP] - [RDP] \qquad (31)$$

$$[UDP] \text{ (g/kgDM)} = [CP] - \{[QDP] + [SDP]\} \qquad (32)$$

$$[DUP] \text{ (g/kgDM)} = 0.9\{[UDP] - 6.25[ADIN]\} \qquad (33)$$

Microbial yield   y (gMCP/MJ FME) = $7.0 + 6.0\{1 - e^{(-0.35L)}\}$ (34)

$$MCP \text{ (g/d)} = ERDP \text{ (g/d)} \qquad (35)$$

$$MCP \text{ (g/d)} (\leq ERDP) = FME \text{ (MJ/d)} \times y \text{ (gMCP/MJ FME)} \qquad (36)$$

Energy retention E_f (MJ/d) = $C4(EV_g \times \Delta W)$ (37)

$$R = E_f/E_m \qquad (38)$$

Maintenance, general	M_m (MJ/d) = (F + A)/k_m	(39)

Fasting metabolism:

Cattle	F (MJ/d) = C1$\{0.53(W/1.08)^{0.67}\}$	(40)
Sheep < 1 year	F (MJ/d) = C1$\{0.25(W/1.08)^{0.75}\}$	(41)
> 1 year	F (MJ/d) = C1$\{0.23(W/1.08)^{0.75}\}$	(42)
Goats	F (MJ/d) = $0.315W^{0.75}$	(43)

Activity allowances:

Housed dairy cows	A (kJ/d) = 9.47W	(44)
Housed beef cattle	A (kJ/d) = 7.08W	(45)
Housed lactating ewes	A (kJ/d) = 9.6W	(46)
Housed pregnant ewes	A (kJ/d) = 5.44W	(47)
Lowland ewes outdoors	A (kJ/d) = 10.7W	(48)
Hill ewes grazing	A (kJ/d) = 23.9W	(49)
Housed growing lambs	A (kJ/d) = 6.7W	(50)
Lowland goats	A (kJ/d) = 18.7W	(51)
Hill goats	A (kJ/d) = 23.9W	(52)

Energy value of milk:
Dairy cows

$$[EV_l] \text{ (MJ/kg)} = 0.0384[BF] + 0.0223[P] + 0.0199[La] - 0.108 \quad (53)$$

$$[EV_l] \text{ (MJ/kg)} = 0.0376[BF] + 0.0209[P] + 0.948 \quad (54)$$

$$[EV_l] \text{ (MJ/kg)} = 0.0406[BF] + 1.509 \quad (55)$$

$$M_l \text{ (MJ/d)} = (Y \times [EV_l])/k_l \quad (56)$$

Ewes $[EV_l]$ (MJ/kg) = 0.0328[BF] + 0.0025d + 2.2033 $\quad$ (57)

$$[EV_l] \text{ (MJ/kg)} = 0.04194[BF] + 0.01585[P] + 0.2141[La] \quad (58)$$

Goats, Anglo-Nubian M_l (MJ/d) = (Y x 3.355)/k_l (59)

Saanen/Toggenburg M_l (MJ/d) = (Y x 2.835)/k_l (60)

Energy value of liveweight gains:

Cattle

$$[EV_g] \text{ (MJ/kg)} = \frac{C2(4.1 + 0.0332W - 0.000009W^2)}{(1 - C3 \times 0.1475\Delta W)} \quad (61)$$

$$E_f \text{ (MJ/d)} = (\Delta W \times [EV_g]) \quad (62)$$

Lambs, entire males $[EV_g]$ (MJ/kg) = 2.5 + 0.35W   (63)

    castrates $[EV_g]$ (MJ/kg) = 4.4 + 0.32W   (64)

    females $[EV_g]$ (MJ/kg) = 2.1 + 0.45W   (65)

Goat kids E_g (MJ/kg) = 4.972W + 0.1637W^2 (66)

    $[EV_g]$ (MJ/kg) = 4.972 + 0.3274W   (67)

Goat fibre,   Cashmere E_w = 0.08MJ/d   (68)

    Angora E_w = 0.25MJ/d   (69)

Net energy of gravid foetus:

Pregnant cattle $\log_{10}(E_t) = 151.665 - 151.64e^{-0.0000576t}$ (70)

    E_c (MJ/d) = 0.025W_c(E_t x 0.0201$e^{-0.0000576t}$)   (71)

Calf birthweight W_c (kg) = ($W_m^{0.73}$ - 28.89)/2.064   (72)

Pregnant ewes $\log_{10}(E_t) = 3.322 - 4.979e^{-0.00643t}$ (73)

    E_c (MJ/d) = 0.25W_0(E_t x 0.07372$e^{-0.00643t}$)   (74)

Energy value of liveweight change:

Cows $[EV_g]$ for liveweight change in lactating cows = 19MJ/kg   (75)

    ME from liveweight loss in lactating cows = (16/k_l)MJ/kg   (76)

Ewes $[EV_g]$ for liveweight gain in lactating ewes = 23.85MJ/kg   (77)

ME from liveweight loss in lactating ewes $= (20.0/k_l)$MJ/kg $\quad$ (78)

MP requirements: general

$$\text{MPR (g/d)} = NP_b/k_{nb} + NP_d/k_{nd} + NP_l/k_{nl} + NP_c/k_{nc} + NP_g/k_{ng} + NP_w/k_{nw}$$
$$(79)$$

Maintenance:

Cattle and goats $\quad\quad NP_m$ (g/d) $= NP_b + NP_d$ $(NP_b = BEN)$ $\quad$ (80)

$$MP_m \text{ (g/d)} = 2.30W^{0.75} \quad\quad (81)$$

Ewes $\quad\quad\quad\quad\quad MP_m$ (g/d) $= 2.1875W^{0.75} + 20.4$ $\quad$ (82)

Lambs $\quad\quad\quad\quad\quad MP_m$ (g/d) $= 2.1875W^{0.75}$ $\quad$ (83)

Cattle, sheep and goats $\quad$ BEN (gN/d) $= 0.35W^{0.75}$ $\quad$ (84)

$$MP_b \text{ (g/d)} = 2.1875W^{0.75} \quad\quad (85)$$

Cattle and goats $\quad\quad MP_d$ (g/d) $= 0.1125W^{0.75}$ $\quad$ (86)

Lactation MP$_l$ (g/kg milk) $= 1.471$ x (True protein content of milk) $\quad$ (87)

Dairy and suckler cows $\quad MP_l$ (g/kg milk) $= 13.57P\%$ $\quad$ (88)

Ewes $\quad\quad\quad\quad\quad MP_l$ (g/kg milk) $= 71.9$ $\quad$ (89)

Goats, Anglo-Nubian $\quad MP_l$ (g/kg milk) $= 47.7$ $\quad$ (90)

$\quad\quad Saanen/Toggenburg$ MP$_l$ (g/kg milk) $= 38.4$ $\quad$ (91)

Liveweight gain:
Cattle

$$NP_f \text{ (g/d)} = \Delta W\{168.07 - 0.16869W + 0.0001633W^2\} \text{ x } \{1.12 - 0.1223\Delta W\} \quad (92)$$

$$MP_f \text{ (g/d)} = C6\{168.07 - 0.16869W + 0.0001633W^2\}$$

$$\text{x } \{1.12 - 0.1223\Delta W\} \text{ x } 1.695\Delta W \quad\quad (93)$$

Sheep:
male/castrate $\quad\quad NP_f$ (g/d) $= \Delta W(160.4 - 1.22W + 0.0105W^2)$ $\quad$ (94)

female $\quad\quad\quad\quad NP_f$ (g/d) $= \Delta W(156.1 - 1.94W + 0.0173W^2)$ $\quad$ (95)

male/castrate $\quad\quad MP_f$ (g/d) $= 1.695\Delta W(160.4 - 1.22W + 0.0105W^2)$ $\quad$ (96)

female MP_f (g/d) = $1.695 \Delta W(156.1 - 1.94W + 0.0173W^2)$ (97)

male/castrate
$$MP_f + MP_w \text{ (g/d)} = \Delta W(334 - 2.54W + 0.022W^2) + 11.5 \qquad (98)$$

female $MP_f + MP_w$ (g/d) = $\Delta W(325 - 4.03W + 0.036W^2) + 11.5$ (99)

Goats NP_f (g) = $157.22W - 0.347W^2$ (100)

$[NP_f]$ (g/kgΔW) = $157.22 - 0.694W$ (101)

MP_f (g/kgΔW) = $266 - 1.18W$ (102)

Lambs' fleece NP_w (g/d) = $3 + 0.1 \times NP_g$ (103)

MP_w (g/d) = $11.54 + 0.3846 \times NP_f$ (104)

Ewes' fleece MP_w (g/d) = 20.4 (105)

Goat fibre,   *Cashmere* MP_w (g/d) = 13.6 (106)

Angora MP_w (g/d) = 38.5 (107)

Pregnant cattle NP_c (g/d) = $TP_t \times 34.37e^{-0.00262t}$ (108)

$\log_{10}(TP_t) = 3.707 - 5.698e^{-0.00262t}$ (109)

MP_c (g/d) = $1.01W_c\{TP_t \times e^{-0.00262t}\}$ (110)

Pregnant ewes NP_c (g/d) = $TP_t \times 0.06744e^{-0.00601t}$ (111)

$\log_{10}(TP_t) = 4.928 - 4.873e^{-0.00601t}$ (112)

MP_c (g/d) = $0.25W_i\{0.079TP_t \times e^{-0.00601t}\}$ (113)

Liveweight change whilst lactating:

Cows MP_g (g/kg) liveweight gain in cows = 233   (114)

MP_g (g/kg) liveweight loss in cows = 138   (115)

Ewes MP_g (g/kg) liveweight loss in ewes = 119   (116)

MP_g (g/kg) liveweight gain in ewes = 140   (117)

Feed evaluation

Fresh grass	ME (MJ/kgDM)	$= 16.20 - 0.0185$[MADF]	(118)
		$= 3.24 + 0.0111$[NCD]	(119)
		$= -0.46 + 0.0170$[IVD]	(120)

Grass hays	ME (MJ/kgDM)	$= 15.86 - 0.0189$[MADF]	(121)
field cured		$= 4.28 + 0.0087$[NCD]	(122)
		$= 2.67 + 0.0110$[IVD]	(123)
barn dried	ME (MJ/kgDM)	$= 1.80 + 0.0132$[NCD]	(124)
		$= 0.61 + 0.0148$[IVD]	(125)

Dried grass	ME (MJ/kgDM)	$= 16.90 - 0.0224$[MADF]	(126)
(high temperature)		$= -0.59 + 0.0154$[NCD]	(127)
		$= -1.82 + 0.0195$[IVD]	(128)

Dried lucerne	ME (MJ/kgDM)	$= 13.90 - 0.0164$[MADF]	(129)
(high temperature)		$= -0.61 + 0.0151$[NCD]	(130)
		$= -0.49 + 0.0163$[IVD]	(131)

Grass silage	ME (MJ/kgDM)	$= 15.0 - 0.0140$[MADF]	(132)
		$= 5.45 + 0.0085$[NCD]	(133)
(also by NIR)		$= 2.91 + 0.0120$[IVD]	(134)

| *Maize silage* | ME (MJ/kgDM) | $= 13.38 - 0.0113$[MADF] | (135) |
| (also by NIR) | | $= 3.62 + 0.0100$[NCD] | (136) |

| *Cereal straws* | ME (MJ/kgDM) | $= 0.53 + 0.0142$[IVD] | (137) |

| " *ammoniated* | ME (MJ/kgDM) | $= 2.24 + 0.0098$[IVD] | (138) |

| " *alkali treated* | ME (MJ/kgDM) | $= 1.62 + 0.0121$[NCD] | (139) |
| (also by NIR) | | $= 2.54 + 0.0093$[IVD] | (140) |

| *Compound feeds* | ME (MJ/kgDM) | $= 0.0140$[NCDG] $+ 0.025$[EE] | (141) |

| | ME (MJ/kgDM) | $= 0.0157$[DOMD] | (142) |

Grass silage only $\quad$ [DOMD$_\text{o}$] (g/kg) $= $ [OMD](1000 - Total ash)/1000 $\quad$ (143)

$$[\text{DOMD}_\text{c}] \ (\text{g/kg}) = 1000 - \{(1000 - [\text{DOMD}_\text{o}]) \times [\text{ODM}]/[\text{CDM}]\} \quad (144)$$

$$[\text{CDM}] \ (\text{g/kg}) = 0.99[\text{ODM}] + 18.2 \quad (145)$$

$$[\text{ME}] \ (\text{MJ/kgCDM}) = 0.16[\text{DOMD}_\text{c}] \quad (146)$$

Degradability parameters $a = 0.06 + 0.61 \times \text{buffer solubility}$ (147)

$$a + b + u = 1 \qquad (148)$$

$$c = r(a - dg_8)/(dg_8 - a - b) \qquad (149)$$

Grass silage [FME] $[\text{FME}] \ (\text{MJ/kgDM}) = 0.90[\text{ME}] - [\text{ME}_{fat}]$ (150)

Brewery byproducts $[\text{FME}] \ (\text{MJ/kgDM}) = 0.95[\text{ME}] - [\text{ME}_{fat}]$ (151)

Grass silage [FME] only

$[\text{FME}] \ (\text{MJ/kgDM}) = [\text{ME}](0.467 + 0.00136[\text{ODM}] - 0.00000115[\text{ODM}]^2)$
(152)

Cattle: prediction of gain $\Delta W \ (\text{kg/d}) = E_f/(X + 0.1475E_g)$ (153)

  where $X = C2(4.1 + 0.0332W - 0.000009W^2)$ *taken from equation (61)*

Dry matter intake $\text{DMI} \ (\text{kg/d}) = \text{MER}/(\text{M/D})$ (154)

Dairy cattle $\text{DMI} \ (\text{kg/d}) = 0.076 + 0.404C + 0.013W - 0.129n$
$\qquad\qquad\qquad + 4.12\log_{10}(n) + 0.14Y$ (155)

$\text{SDMI} \ (\text{kg/d}) = -3.74 - 0.387C + 1.486(F+P) + 0.0066W_n + 0.0136[\text{DOMD}]$
(156)

$\text{SDMI} \ (\text{kg/d}) = -3.74 - 0.387C + 0.1055Y + 0.0066W_n + 0.0136[\text{DOMD}]$
(157)

$I \ (\text{g/kgW}^{0.75}) = 0.103[\text{DM}] + 0.0516[\text{DOMD}] - 0.05N_a + 45$ (158)

$\text{SDMI} \ (\text{kg/d}) = (1.068xI - 0.00247(IxC) - 0.00337C^2 - 10.9)W^{0.75}/1000$
$\qquad\qquad\qquad + 0.00175Y^2$ (159)

Lactation curve for dairy cattle

$$Y \ (\text{kg/d}) = \exp\{a - bt1(1 + kt1) + ct1^2 + d/t\} \qquad (160)$$

$$Y \ (\text{kg/d}) = \exp\{3.25 - 0.5t1(1 + 0.39t1) - 0.86/t\} \qquad (161)$$

Dry matter intake:
Pregnant cattle

$$\text{SDMI} \ (\text{kg/d}) = \{0.0003111[\text{DOMD}] - 0.00478C - 0.1102\}W^{0.75} \quad (162)$$

Beef cattle

SDMI (kg/d) =

$$W^{0.75}(24.96 - 539.7C + 0.108[TDM] - 0.0264N_a + 0.0458[DOMD])/1000 \tag{163}$$

coarse diets $\quad$ TDMI $(g/kgW^{0.75}) = 24.1 + 106.5q_m + 0.37C\%$ $\quad$ (164)

fine diets $\quad$ TDMI $(g/kgW^{0.75}) = 116.8 - 46.6q_m$ $\qquad$ (165)

Lactation curve for suckler cows $\quad$ Y $(kg/d) = 8.0n^{0.121} \times e^{-0.0048n}$ $\quad$ (166)

Dry matter intake:
Pregnant ewes, hay

HDMI $(kg/d) = C(1.9 - 0.076T - 0.002033[DOMD]) + 0.002444[DOMD]$
$\qquad - 0.09565LS + 0.01891W_8 - 1.44$ $\qquad$ (167)

silage $\quad$ I $(g/kgW) = 0.202[DOMD] - 0.0905W - 0.0273N_a + 11.62$ $\quad$ (168)

$\qquad$ SDMI $(kg/d) = 0.001W\{0.946 \times I - 0.204(C \times I) + 0.569\}$ $\qquad$ (169)

Lactating ewes, hay

$\qquad$ TDMI $(kg/d) = 0.001W\{I - 0.0691(I \times C) + 2.027C\}$ $\qquad$ (170)

$\qquad\qquad$ I $(g/kgW) = 0.0481[DOMD] - 5.25$ $\qquad$ (171)

$\qquad\qquad$ TDMI $(kg/d) = 0.028W$ $\qquad$ (172)

silage TDMI $(kg/d) = 0.001W\{0.946 \times I - 0.0204(I \times C) + 0.65 + C\}$ (173)

$\qquad\qquad$ I $(g/kgW) = 0.0232[DOMD] - 0.1041W - 0.0314N_a + 13.36$ $\quad$ (174)

$\qquad\qquad$ TDMI $(kg/d) = 0.026W$ $\qquad$ (175)

Lambs, coarse diets

$\qquad\qquad$ TDMI $(kg/d) = \{104.7q_m + 0.307W - 15.0\}W^{0.75}/1000$ $\qquad$ (176)

fine diets $\quad$ TDMI $(kg/d) = \{150.3 - 78q_m - 0.408W\}W^{0.75}/1000$ $\qquad$ (177)

Lambs, silage only $\qquad$ SDMI $(kg/d) = 0.046W^{0.75}$ $\qquad$ (178)

Lactating goats

$$\text{DMI (g/d)} = 423.2Y + 27.8EBW^{0.75} + 440_\Delta W + 6.75F\% \qquad (179)$$

$$\text{DMI (kg/d)} = 0.42Y + 0.024W^{0.75} + 0.4_\Delta W + 0.7Fp \qquad (180)$$

$$\text{DMI (kg/d)} = 0.062W^{0.75} + 0.0305Y \qquad (181)$$

Lactation curve for goats

$$Y \text{ (kg/d)} = 3.47\exp\{-0.618(1 + t1/2)t1 - 0.0707t1^2 - 1.01t\} \qquad (182)$$

Dry matter intake:

Adult goats $$\text{DMI (kg/d)} = \{130.9q_m + 0.384W - 18.75\}W^{0.75}/1000 \qquad (183)$$

Pregnant goats $$\text{DMI (kg/d)} = 0.53 + 0.0135W \qquad (184)$$

Subject Index